AF443187

Phosphoric Anhydride

Phosphoric Anhydride

Structure, Chemistry and Applications

Dmitry A. Efremov

Burlington Chemical Co., North Carolina, USA

Pavel M. Zavlin

St Petersburg Institute of Cinema and Television, Russia

John C. Tebby

Staffordshire University, UK

JOHN WILEY & SONS

Chichester · New York · Weinheim · Brisbane · Singapore · Toronto

Other Wiley Editorial Offices

John Wiley & Sons, Inc., 605 Third Avenue,
New York, NY 10158–0012, USA

WILEY–VCH Verlag GmbH, Pappelallee 3,
D-69469 Weinheim, Germany

Jacaranda Wiley Ltd, 33 Park Road, Milton,
Queensland 4064, Australia

John Wiley & Sons (Asia) Pte Ltd, 2 Clementi Loop #02-01,
Jin Xing Distripark, Singapore 129809

John Wiley & Sons (Canada) Ltd, 22 Worcester Road,
Rexdale, Ontario M9W 1L1, Canada

Library of Congress Cataloging-in-Publication Data

Efremov, Dmitry A.
   Phosphoric anhydride : structure, chemistry and applications /
Dmitry A. Efremov, Pavel M. Zavlin, John C. Tebby.
     p. cm.
   Includes bibliographical references and index.
   ISBN 0-471-98508-2 (hardcover : alk. paper)
   1. Phosphoric anhydride. I. Zavlin, Pavel M. II. Tebby, John
C., 1933– . III. Title.
QD304.A2E26 1998   98–20534
546'.71222 — dc21   CIP

British Library Cataloguing in Publication Data

A catalogue record for this book is available from the British Library

ISBN 0-471-98508-2

Typeset in 10/12pt Times by Dobbie Typesetting Limited, Tavistock, Devon.
Printed and bound in Great Britain by Biddles Ltd., Guildford and King's Lynn.
This book is printed on acid-free paper responsibly manufactured from sustainable forestry in which at least two trees are planted for each one used for paper production.

Contents

Preface

This book is written for specialists and students in organic, organophosphorus, biological, inorganic and industrial chemistry.

Phosphoric anhydride is a unique reagent in organic synthesis and is involved in the reactions of dehydration, dealcoholysis, phosphorylation, condensation, rearrangement, catalysis, and many others.

Phosphoric anhydride presents a straightforward synthetic approach to acidic phosphates that in many cases is difficult to achieve by other methods.

Phosphoric anhydride is of particular importance in biological chemistry because it is quite probable that phosphoric anhydride and/or its derivatives were the primary phosphorylating agents of organic compounds on the Primitive Earth. If this is true it means that the natural phosphorylating agent of the majority of biological systems is phosphoric anhydride.

Strictly speaking the chemistry of phosphoric anhydride unites inorganic and organic chemistry because nearly all steps of the reactions of phosphoric anhydride with organic compounds result in the formation of new P–Element bonds followed by the disintegration of phosphoric anhydride's residual polyphosphate chain.

Phosphoric anhydride gives rise to numerous organic and inorganic phosphates that are all around us in our textiles, food and drinks, paper, plastics, and so on. A tremendous variety of important phosphates are manufactured by hundreds of chemical companies, and we hope that acting and potential manufacturers of phosphates will find much useful information for this fruitful business.

This book will appeal to postgraduate students and scientists because it shows not only the achievements in the chemistry of phosphoric anhydride but also its huge potential. For instance, many reactions with phosphoric anhydride were carried out decades ago and have never been refined with modern experimental and theoretical methods. So, many findings are waiting for a scrutinizer. Are they not?

Undergraduate students will discover new valuable information about 'the familiar' compound, its chemical activity, and the properties of the products derived from it. Keeping in mind chemistry undergraduate and postgraduate students we have presented most parts of the book on the principle of the historical development of phosphoric anhydride.

Introduction

Both inorganic and organic derivatives of phosphorus are an integral part of nature. The recent achievements of biology and chemistry, a better understanding of the processes in biological and chemical systems helps us to realize the unique role of phosphorus in the evolution and functioning of life. Many biological processes involve the principal step of phosphoryl group transport in a substrate. Naturally the reactions of phosphorylation of various nucleophilic compounds are of particular interest for chemists, such phosphorylating agents as phosphorus halides, esters, amides and amidoesters of phosphoric acid have been studied in depth. In this respect phosphoric anhydride used to be the Cinderella of bioorganic and organic studies. Phosphoric anhydride was and is widely used as a dehydrating agent and only a limited number of researchers were curious enough to scrutinize these complicated processes. Nevertheless several magic reactions were discovered which involved the active participation of phosphoric anhydride. However, research devoted to the reactivity of phosphoric anhydride published in prominent journals, patents and elsewhere has never been collated, which limits the overall knowledge of phosphoric anhydride and retards its application. We, therefore, undertook the responsibility to summarize the corresponding literature data and apply our own rich experience in predominantly nucleophilic phosphoric anhydride chemical processes. In such reactions the nucleophilic susceptibility of phosphorus in phosphoric anhydride is strengthened by the protonating activity of protic nucleophiles (H–Nu).

We consider such an approach as a general primary step of phosphoric anhydride reactions with O–H, N–H, S–H, C–H and other nucleophiles that are presented in the following Chapters 3–6.

Writing this book provoked a strong temptation to include as much information as possible concerning phosphoric anhydride. Nevertheless we narrowed down the list of references to about 850 of the most important publications devoted to phosphoric anhydride or relevant to it.

Please do not be misled by a relatively small Section 6.3, 'Preparation and application of inorganic compounds derived from phosphoric anhydride'. Many aspects of this topic can be found in the other parts of this book in order to express phosphoric anhydride chemistry more solidly. For example the reaction of phosphoric anhydride with ammonia (Chapter 3, Section 3.1) prefaces the reactions with amines (Chapter 3, Section 3.2) and other nitrogen-containing compounds, the processes with PCl_3, PCl_5, $POCl_3$ and H_3PO_4 are presented mostly in Chapter 3, Section 4. We have limited the number of examples of the most familiar inorganic reactions of phosphoric anhydride in favour of processes of less common knowledge.

NOTE FROM DMITRY A. EFREMOV

When this book was ready for publication, our colleague, brilliant scientist, lively and friendly person Professor Pavel M. Zavlin passed away.

CHAPTER 1

Preparation of Phosphoric Anhydride

Phosphoric anhydride (P_4O_{10}) is prepared generally by either of two methods: burning of phosphorus or by ignition of calcium phosphate with silicon oxide. Since its discovery by Robert Boyle in 1680 the interaction between phosphorus and oxygen has been thoroughly explored. The method of direct oxidation of phosphorus is used widely in the industry [1]. Phosphorus vapour is passed through specific burners in a stream of dry oxygen or air. The lack of oxygen or air in the stream can result in the formation of significant quantities of lower phosphorus oxides (P_4O_6 and P_4O_8). The resulting phosphoric anhydride vapour is condensed in cameras into white powder. Sublimation in dry oxygen then helps to purify the product. The traces of the lower oxides can be further oxidized into phosphoric anhydride by heating the oxides to 175–200°C in the stream of oxygen or by treatment with ozone. Phosphoric anhydride produced in such a way is characterized by: specific density 2.39 g/cm^3 (20°C) [2]; melting point 423°C (quick heating under atmospheric pressure), and 563°C (under high pressure); temperature of sublimation 347°C (under normal pressure); and boiling point 864°C (under normal pressure) [3]. Specific density of P_4O_{10} condensed at different temperatures varies within the range between 2.330 g/cm^3 (-78°C) [4] and 2.228 g/cm^3 (20°C) [4]. The bulk phosphoric anhydride is produced with specific density 0.77–1.39 g/cm^3.

Muenow and co-workers [5] determined by means of mass spectrometry that at temperatures below 150°C the equilibrium vapour of commonly available phosphoric anhydride consists predominantly of P_4O_{10} with insignificant impurity of P_4O_9. At the temperature range between 175 and 350°C phosphoric anhydride vapour contains P_4O_{10} and P_4O_9, whereas at temperatures above 400°C the presence of P_4O_8 was registered as well.

As noted [6] the complicated reaction of phosphorus with oxygen proceeds via the formation of a wide variety of phosphorus oxides. Matrix reaction of phosphorus with oxygen (Schemes 1–3) gave phosphorus pentoxide as one of the products [6].

$$P_4 + O \longrightarrow [P_4O] \longrightarrow P_2 + [P_2O]$$
$$[P_2O] \longrightarrow P + PO \tag{1}$$
$$PO + O \longrightarrow PO_2 \tag{2}$$
$$PO_2 + PO + O_2 \longrightarrow P_2O_5 \tag{3}$$

For the monomeric (P_2O_5) versus dimeric (P_4O_{10}) structure of phosphoric anhydride see Section 2.6.

Mielke and Andrews [7] studied the reaction of P_4O_6 with oxygen isotopes (^{16}O and ^{18}O) (Scheme 4) generated in the photolysis of ozone. The products were trapped in solid argon and examined using IR spectroscopy (see also Section 2.6).

$$nP_4O_6 \xrightarrow{\ ^{16}O(^{18}O)\ } P_4O_7 + P_4O_8 + P_4O_{10} + P_4O_6^{18}O + P_4O_6^{18}O_2 + P_4O_6^{18}O_4 \tag{4}$$

Van Wazer and co-workers [8, 9] claimed that the anionic form of phosphoric anhydride 'cage' was formed as a product of 'the reverse decondensation reaction' in the process of the interaction between thoroughly dehydrated (crystalline) orthophosphoric acid and an excess of tetramethylurea and/or diisopropylcarbodiimide. The experiments were carefully monitored using phosphorus-31 NMR. According to the spectrometric data obtained in one of the experiments the authors considered the two possible structures of dianion either **(1)** or its anionic dimeric form **(2)**.

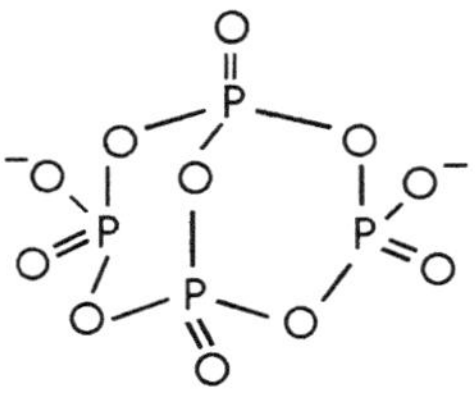

(1) 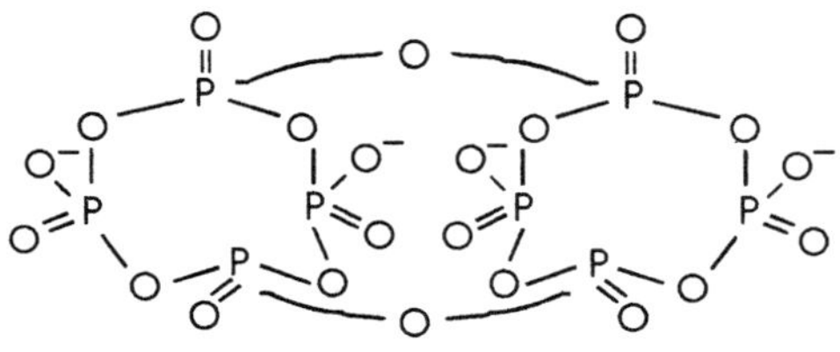   **(2)**

The computer simulation of the spectra for both structures led the authors [9] to favour structure **(2)**. Tricyclic ('cage') 1,5-μ-orthophosphatyltetrametaphosphate anion **(2a)** was also produced [9].

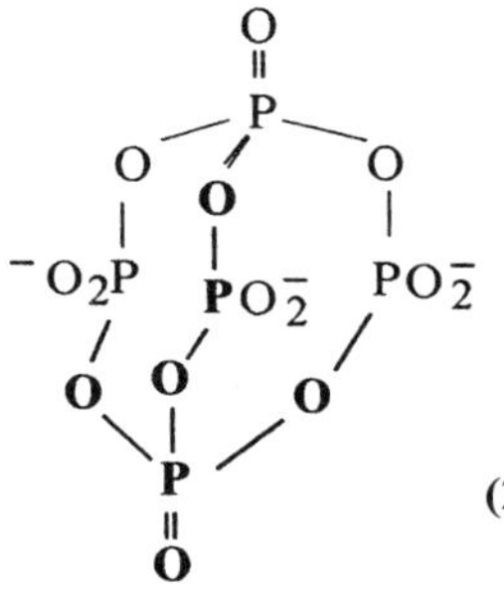

(2a)

On the basis of earlier studies [8] Van Wazer and co-workers went further in their attempts of tetrametaphosphate condensation-cyclization [10] targeting to make the molecule P_4O_{10}. The authors took advantage of dicyclohexylcarbodiimide, with its melting point close to the room temperature and its much lower ability to dissolve the corresponding urea than ultraphosphate molecules [8, 10]. This method of condensation presented an efficient approach to 1,5-μ-oxo-tetrametaphosphate.

The bridging phosphorus atoms in the product were registered in phosphorus-31 NMR by the multiplet in the region $c. -40$ ppm.

The further 'pushing' of 1,5-μ-oxo-tetrametaphosphate condensation-cyclization with an excess of dicyclohexanecarbodiimide resulted in a new pattern of the phosphorous-31 NMR spectra that possessed the signals in the extremely high-field regions: 59.5 ppm (relative area $c. 3.0$) and -64 ppm (relative area $c. 1.0$), which could be assigned to the P_4O_{10}-carbodiimide zwitterionic structure [10] as presented by Scheme 5.

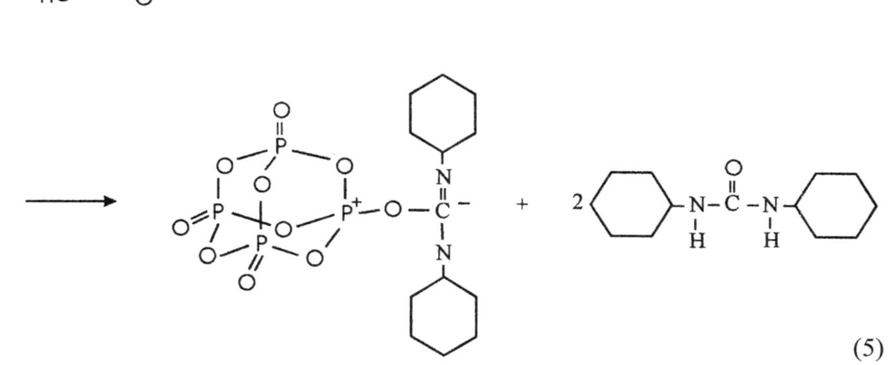

(5)

The value of phosphorus–phosphorus coupling was determined to be $^2J_{P-P}$ 18.0 Hz and the decoupling experiments exhibited no P–H coupling effects. Signal-averaged homonuclear ^{31}P–^{31}P INDOR experiments [10] confirmed the origin of the coupling. Addition of tetramethyl urea to the product resulted in the decrease of the multiplet intensity at -60 ppm and signals of 1,5-μ-oxo-tetrametaphosphate reappeared due to hydrolysis [10]. Thus this study was the first impressive reconstruction of the P_4O_{10} tricyclic structure.

Phosphoric anhydride has also been formed in a partial thermal decomposition of trialkylsilyldifluorophosphates [11] (Scheme 6).

$$6R_3SiOP(O)F_2 \longrightarrow 6R_3SiF + 2POF_3 + P_4O_{10} \tag{6}$$

CHAPTER 2

Structure and Physico-chemical Properties of Phosphoric Anhydride

1 CRYSTALLINE STRUCTURE

Earlier studies of a phase diagram of phosphoric anhydride [12] provided some evidence of the existence of three crystalline forms of phosphoric anhydride. Temperature and pressure significantly influenced which crystal structures were formed [13]. A first in-depth study of the crystalline forms of phosphoric anhydride was carried out by de Decker [14, 15]. He sublimed phosphoric anhydride in a sealed tube and used X-ray analysis to determine a cell consisting of eight P_2O_5 units, the main element of which is a deformed tetrahedral PO_4.

In 1943 Hill *et al.* [16] described three crystalline modifications of phosphoric anhydride; one of which the so-called 'commercial phosphorus pentoxide' contained mainly the metastable crystalline rhombohedral (α) form of the hexagonal system. The rate of conversion of this form into the other two metastable orthorhombic and stable tetragonal forms under normal conditions was immeasurably slow.

The conversion of rhombohedral form into orthorhombic form at 180°C lasted for *c.* 1440 h and at 378°C it took about 1 h.

MacGillavry and co-workers [17] were cautious concerning the structure of the third crystalline form of phosphoric anhydride described by Hill *et al.* [16]. It had been formed by the firm adhesion of phosphoric anhydride to the container's glass inner surface and they were aware that this may have been followed by a chemical reaction with the glass. For confirmation the authors [17] carried out an X-ray study which revealed a corrugated sheets structure (Compound **3**) characterized by the following data: space group Pnam–D_{2h}^{16}; a=9.23 Å, b=7.18 Å, c=4.94 Å.

'Corrugated sheets'   (3)

This structure was later refined by Cruickshank [18]. In the publications [16, 17] the molecular structure of phosphoric anhydride is presented as $(P_2O_5)_\infty$. The authors [16] considered the links in the corrugated sheets of phosphoric anhydride to be similar to those in vanadium pentoxide. In all other respects the smaller size of the phosphorus atom resulted in a less deformed tetrahedral structure, more isotropic binding in the sheets and more densed packing.

Thirty-seven years after de Decker's publication [17], Jansen and Luer [19] refined those characteristics using diffractometry and Guinier powder data for phosphoric anhydride sublimed in argon. The rhombododecahedral structure of P_4O_{10} studied by the authors [19] belonged to the space group R3c with the lattice constants a=10.3035 Å and c=13.5102 Å.

Some aspects of the crystalline structure of phosphoric anhydride are discussed in Section 3 in connection with its stereochemical studies.

2 MOLECULAR FORMULA OF PHOSPHORIC ANHYDRIDE

In 1896 Tilden and Barnett published a paper in the *Journal of the Chemical Society* [20] presenting the results of their study of phosphoric anhydride vapour density. The main objective for the investigation was the contradictory fact that phosphorus lower oxide (P_4O_6), arsenic and antimony oxides contained four metalloid atoms in a molecule with the only exception being phosphoric anhydride whose formula was considered to be P_2O_5 [21]. Prior to the experiments the authors purified phosphoric anhydride to remove the lower oxides by sublimation in dry oxygen above heated porous platinum. In all six experiments described, phosphoric anhydride vapour density correlated with a molecular weight between 307 and 370. This was in better agreement with the formula P_4O_{10} (MW 284) rather than P_2O_5 (MW 142). The discussions concerning the molecular formula of phosphoric anhydride lasted until the 1960s. So the publication [20] might be considered as the starting point of the study of phosphoric anhydride. It is noteworthy that in the same publication [20] Tilden and Barnett expressed the idea that the higher phosphorus sulphide possessed the structure that corresponds to the formula P_4S_{10} and all prior experiments that had determined this formula were caused by the dissociation of the higher sulphide at the temperature of sublimation into P_4S_6 and S_4.

Tilden and Barnett [20] proposed a rigid polycyclic structural formula (4) of phosphoric anhydride that contained 10 bridge oxygen atoms. Two possible stereostructures (5, 6) of P_4O_{10} were discussed by Khodakov [22] who supported strongly structure (6).

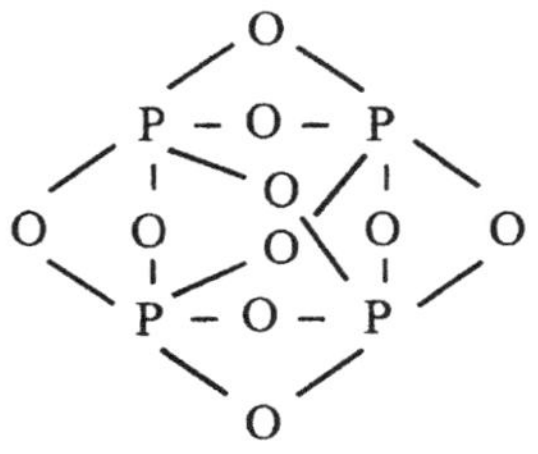

(4) 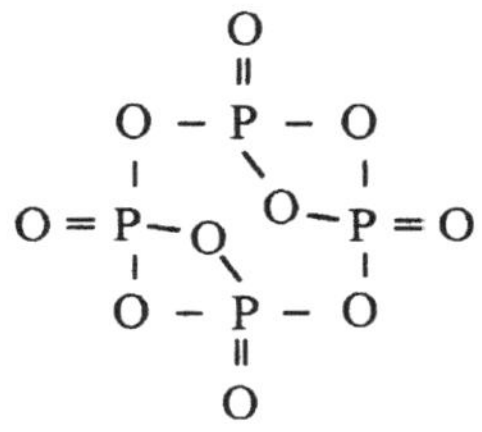　　　　　　　(5) 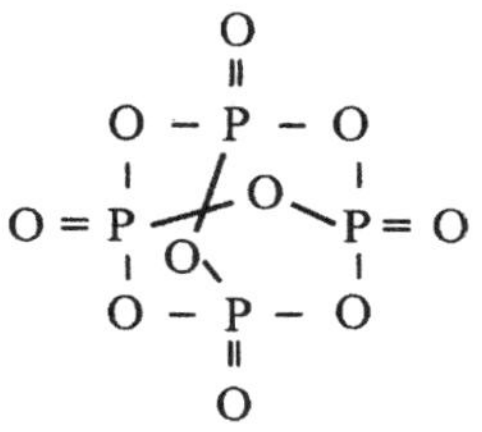　　　　　　　(6)

The naming of the compound P_4O_{10} has been a problem for many chemists. The term 'phosphorus pentoxide' and the formula P_2O_5 are used for phosphoric anhydride (P_4O_{10}) in a good number of recent publications including scientific publications, technological reports and textbooks. In *Chemical Abstracts* it is listed under 'Phosphorus Oxide, P_2O_5', in other texts it maintains its original name of phosphorus pentoxide. It is frequently called phosphoric anhydride or phosphoric oxide owing to its hexahydrate relationship with phosphoric acid. While its full systematic name would be very complex and clumsy, the alternative names 2,4,6,8,10-hexoxy-1,3,5,7-tetraphosphacyclodecane-1,3,5,7-tetroxide or 1,5-μ-oxo-3,7-μ-oxo-tetrametaphosphate do not seem to have been accepted.

3 ELECTRON DIFFRACTION AND X-RAY STUDIES

The earliest attempt to study P_4O_{10} by means of electron diffraction was made by Maxwell, Hendricks and Deming [23] in 1937. The structures of tetraphosphorus hexoxide, tetraphosphorus octaoxide and tetraarsenic hexoxide were investigated and comparatively analysed. The stereochemical parameters determined for the latter two compounds correlated well enough with the symmetry of the point group T_d. The angle P–O–P in tetraphosphorus hexoxide had the value $128.5\pm1.5°$ and the P–O bond length 1.67 ± 0.05 Å. For its analogue tetraarsenic hexoxide it was much more difficult to determine the molecular dimensions. Initially, the scientists [23] assumed that phosphoric anhydride had the same molecular symmetry belonging to the point group T_d, that is the additional four external oxygen atoms did not distort the tricyclic skeleton of the type P_4O_6. However, the authors [23] produced electron diffraction results that did not correlate well with the symmetry of the point group T_d. In fact the best agreement for relating the parameters of P_4O_{10} and P_4O_6 was observed for the structure of the former compound with the exocyclic P=O bond (1.60 Å) which was improbable. The natural consideration of the lower type of symmetry for phosphoric anhydride also had to be rejected because the overall arrangement of the four phosphorus atoms in it (Table 2.1) was similar to that in tetraphosphorus hexoxide. So the resulting structure of phosphoric anhydride remained uncertain.

The authors of the publication [23] also described an interesting observation: the transformation of the yellow phosphorus lattice consisting of small groups of atoms with the covalent angle P–P–P $c.\,60°$ into the highly developed multiatom lattice of black phosphorus involving a widening of the corresponding angle P–P–P to $c.\,102°$. The latter data is quite close to valence angles in phosphorus trichloride, phosphorus oxychloride and tetraphosphorus hexoxide.

Hampson and Stosick [24] attributed the failure to determine the stereostructure of phosphoric anhydride [23] to the presence of an extraordinary short P–O bond. The publication was based on electron diffraction experiments in which they listed not only the stereochemical data for P_4O_{10} (Table 2.1) but presented the three-dimensional structure of a phosphoric anhydride molecule (**7**). The model generally accepted up to now is a tricyclic adamantan-like structure with four exceptionally short P=O bonds, 1.390 Å. Besides phosphoric anhydride the authors [24] studied the structures of tetraphosphorus hexoxide, tetraarsenic hexoxide and hexamethylene tetramine which was very important for the correlation analysis of the experimental data. This work is of particular value because of the study of the same compounds with the help of two different instruments, which was a rare practice in the 1930s. The two sets of data determined with two electron diffraction cameras varied within $\pm0.5\%$. Principally the structures of lower phosphorus and arsenic oxides differed insignificantly taking into consideration the specific features of P and As atoms. Both types of atom exhibited a pronounced tendency in their lower oxides to the formation of smaller than tetrahedral angles. The models worked out for P_4O_6 and As_4O_6 were of the same type. The eventual choice between the models provided data for tetraphosphorus hexoxide which was comparable with the corresponding results described one year earlier by Maxwell, Hendricks and Deming [23]. According to the publication [24] four exocyclic P=O bonds were of insignificant influence upon the skeleton structure of tetraphosphorus hexoxide. The shortened P=O bonds in phosphoric anhydride are explained by the scientists [24] in terms of single–double bond resonance and polar character of the bond.

"The short P–O distance in P_4O_{10} is in accord with the chemical properties of this molecule. Such short bonds are presumably extremely stable, and the thermal stability and resistance to reduction of P_4O_{10} are well known" (Hampson and Stosick, 1938 [24]).

Three years later the X-ray experiments performed with the metastable crystalline form of phosphoric anhydride by de Decker and MacGillavry [15] corroborated the stereochemical data (Table 2.1) and the phosphoric anhydride molecular model of Hampson and Stosick [24]. The authors presented P_4O_{10} as a molecule consisting of the four PO_4 elements that possess the distorted tetrahedral structure (7).

(7)

Phosphoric anhydride electron diffraction data presented in the publication [25] (Table 2.1) supported these models for phosphoric anhydride with some variations. The Russian chemists attributed the variations in their results and the earlier data [24] to incorrect visual determinations of corresponding intensities.

Cruickshank presented a number of interactive papers [18, 26–33] which were devoted to refinement of electronic and stereochemical properties of phosphorus oxides and phosphates. He was the first to discuss the sharing of π character by the P–O bond in phosphoric anhydride [18] and worked out an inverse proportional relationship between the P–O bond length and π-bond order [31].

The stereochemical data of the article [30] differ vividly from those of the other publications (Table 2.1), which Cruickshank explained by inaccuracy of the earlier calculations and methods used. Another distinctive feature of the data under discussion [30] is a significant difference between the similar structural elements: endocyclic P–O bonds, exocyclic P=O bonds, P–O–P and O–P–O angles. The exocyclic P=O bonds presented in the paper [30] are considerably longer than determined by the other investigators and vary considerably (1.41±0.02 Å and 1.51±0.04 Å). In the studies [18, 30] Cruickshank correlated the molecular structure and crystalline packing of phosphoric anhydride. Phosphoric anhydride crystals of the first form belong to the rhombohedral space group R3$_c$ with characteristic values a=7.43 Å and α=87°. The P=O bonds of molecules are orientated along the crystallographic threefold axis of symmetry and a molecule at (0, 0, 0) is twisted around the threefold axis opposite a molecule at (±1/2, ±1/2, ±1/2) by the action of the glide plane symmetry. This close packing achieved by a low symmetrical form is due to '*the roughly tetrahedral envelope of the molecule*' [30]. According to the author [30] such type of crystalline arrangement provokes the shortening of one of the exocyclic P=O bonds in relation to the other external bonds.

A few years later Cruickshank and co-authors [32] made an attempt to achieve more accurate experimental electron diffraction data for the further theoretical study of d and p orbital involvement in the π-bonding in phosphoric anhydride. As a result, different data were presented compared to the earlier publications [18, 30] (Table 2.1). These data [32] were repeatedly quoted in the subsequent publications [19, 33–36]. Besides phosphoric anhydride, another phosphorus oxide with three tetracoordinated and one tricoordinated phosphorus atom was investigated, i.e. tetraphosphorus nonoxide (P_4O_9). The presence of the unshared pair of electrons distorted the whole tetraphosphorus hexoxide 'cage' causing the three types of the bridge P–O bonds to differ significantly: (-O)$_2$ P–O 1.661 Å, (-O)$_2$ P-O — P(O)(O-)$_2$ 1.605 Å, (-O)$_2$(O) P-O — P(O)(O-)$_2$ 1.590 Å. The authors [32] gave a very clever description of phosphoric anhydride molecule stereostructure which is quoted below:

> "*The geometry of the molecule can be described in terms of three imaginary cubes packed one inside the other. The innermost cube (side 2x$_s$) has phosphorus atoms at four of its corners; the outermost cube (side 2x$_t$) has the peripheral axygen atoms, O (per.), at four of its corners; and the central cube (side 2x$_u$) has the bridging oxygen atoms, O (br.), at the centres of its six faces.*"

For such a phosphoric anhydride structure the length of the P=O bond determined (1.429 Å), contrary to the publication [30], was common for all four fragments.

The latest paper after Cruickshank [33] devoted to the stereochemistry of phosphorus oxides is particularly valuable for comparing the similar stereochemical features of the tricyclic 'cage' of P_4O_6 and P_4O_{10}. The difference in P–O bond lengths of two oxides arises from the different valence and coordination states of P atoms. Similar results are presented by other authors [37–42]. The structural features of P_4O_8 presented [33] a possibility to speculate further on the input of p and d orbitals in π-bonding. Continuing from this the variations in the P–O bonds of tricoordinated and pentacoordinated phosphorus in tetraphosphorus octoxide (P_4O_8) are discussed. A certain confusion in the stereochemical characteristics of phosphoric anhydride as well as the other phosphorus oxides can arise from the fact that in some publications data are generated by different experimental methods. For example in the paper [34] there are tables of geometric parameters of P_4O_6, P_4O_7, P_4O_8, P_4O_9, and P_4O_{10} determined partially for the crystalline structures, and in the other cases for molecules in the gaseous state. On the basis of their own experimental data and a literature analysis Jost and Schneider [34] came to the conclusion that the phosphorus oxides' molecular structures belong to one and the same adamantan-like 'cage' configuration of tetraphosphorus hexoxide that varies by the number of exocyclic P=O bonds. It was pointed out that the crystal structures of phosphorus oxides possess similar lattice characteristics. The accuracy of all measurements performed for these highly hydroscopic compounds rely very much upon the methods of their purification and storage.

Jansen and Luer [19] found considerable differences between the similar structural elements of a P_4O_{10} molecule. Diffractometry data confirmed principally the crystal structure described by de Decker and MacGillavry [15]. The authors [19] criticized the wide data variations for the P–O bridging and P=O exocyclic bonds presented by Cruickshank and co-workers [31, 32] which produced a highly distorted 'cage' structure of phosphoric anhydride and other phosphorus oxides.

The corresponding stereochemical parameters of phosphoric anhydride are summarized in Table 2.1.

Besides lower tetraphosphorus oxides, one of the closest structural analogues of phosphoric anhydride is tetraphosphorus decasulfide (P_4S_{10}). It was described in 1843 by Berzelius [43] who presented it in the form of diphosphorus pentasulfide (P_2S_5).

According to X-ray and NMR data the molecule of P_4S_{10} as well as P_4O_{10} is a tricyclic adamantan-like structure; its stereochemical characteristics are presented as follows [44, 45] **(8)**:

Table 2.1   Molecular stereochemical parameters of P_4O_{10}

Bond length, (Å)		Angles(°)				
P–O	P=O	P–O–P	O–P–O	O–P=O	Method	Ref.
1.62±0.02	1.39±0.02	123.5±1	101.5±1	116.5±1	Electron diffraction	[24]
1.62	1.39	123.5	101.5	116.5	X-ray analysis	[15, 40]
1.56	1.40		109.0		Calculation	[22]
1.60	1.40	124.5			Electron diffraction	[25]
1.59/1.66	1.44	123.6/ 125.6	98.1		X-ray analysis	[41]
1.60	1.40	124.5			Calculation	[30]
1.604	1.429	123.5	101.5	116.5	Electron diffraction	[27, 28]
1.575(4)	1.410(2)/ 1.510(4)					[34, 35]
1.588(3)/ 1.597	1.434(4)/ 1.443	122.6(1)/ 123.0(1	101.3(2)/ 102.3(2)	116.2(2)	Diffractometry	[19]
1.604	1.429		101.6		Spectroscopy	[47]
1.606	1.470	129.3	98.0	119.4	Calculations, MNDO/H	[38]
1.583	1.4266	125.15	100.63	117.31	Calculations, COSMIC	Authors
1.689	1.435	132.08	96.40	120.39	Calculations, PM 3	Authors

S
1.91
2.01
P
S
109
S
110
S
S = P
S
P = S
P
S
S
S

(8)

Naturally all bonds in phosphorus thioanhydride are significantly longer than in phosphoric anhydride and the long highly polarized P=S bond is of considerable importance as it could act as a reaction site unlike P_4O_{10} in which the P=O bond is exceedingly short.

4 MOLECULAR MODELLING

The first attempt to calculate the molecular structural parameters for P_4O_{10} was made by Khodakov [22] in 1944. On the basis of calculations the author described P_4O_{10} as a tetrahedral structure consisting of four similar PO_4 fragments that are slightly flattened towards the centre (Table 2.1). Unfortunately the method of calculation is not described in the paper [22]. The model worked out by Khodakov was confirmed later by spectrometric experiments [46, 47].

Some early results of calculations of phosphoric anhydride are presented above in Section 3 and compared with the experimental methods of analysis (electron diffraction and X-ray) of the molecular structure.

Surprisingly enough there are only a few papers published on quantum-chemical calculations of P_4O_{10}. Lohr [37] applied analytical gradients with the GAUSSIAN 86 program and the split-valence basis set 3–21 G* that contained polarization functions for the second-row phosphorus atom only. The calculations focused on the P_2O_5 molecule which the author considered as a C_2-symmetry element of the dimeric oxo-bridged P_4O_{10} (7). The angles in P_2O_5 were determined to be 131.6° (P–O–P) and 136.3° (O=P–O), two P–O bonds and four P=O had practically no variations. Furthermore the two groups O=P–O were shown to be equivalent.

Slivko and Krivoviazov [38] performed their modelling on the basis of MNDO/H approximation. The corresponding data are presented in Tables 2.1, 2.3.

The authors of this book performed the molecular modelling of P_4O_{10} using the molecular mechanics (MM) method (program COSMIC [48, 49]) (Table 2.1) and semiempirical programs AM 1 [50] and PM 3 [51, 52] (Tables 2.1, 2.2). (This is the first presentation of the data.) MM COSMIC was chosen as a very simple and time-saving method that in many cases gives rather satisfactory correlations with the results of higher level calculations [53, 54]. The charges were determined according to LIVERPOOL 2 method. Unfortunately the program AM 1 failed to calculate the structure of P_4O_{10} which is consistent with the results described in the paper [55]. It is noteworthy that the MM data of P_4O_{10} stereostructure compare favourably with the experimental results of electron diffraction [19] whereas PM 3 data correlate better with those of X-ray analysis [41] though the PM 3 program does not deal with molecules in the crystalline state. The similar types of bonds in phosphoric anhydride were determined to be equal.

According to MM COSMIC and PM 3 data the oxygen atoms of the P=O groups make the biggest input in HOMO and the phosphorus atoms make the biggest input in LUMO.

The dipole moment of P_4O_{10} calculated by PM 3 is nearly zero and this correlates with the symmetry of the type T_d.

Table 2.2   The coordinates of atoms in P_4O_{10} calculated by PM 3 [Authors]

Number of atoms	Atom	X	Y	Z
1	P	0.0000	0.0000	0.0000
2	O	1.6880	0.0000	0.0000
3	P	2.8268	1.2474	0.0000
4	O	2.1218	2.1287	1.2576
5	P	0.5307	2.6171	1.5442
6	O	−0.1822	1.1126	1.2591
7	O	−0.7370	−1.2305	−0.0020
8	O	4.2317	0.9574	0.0077
9	O	0.1958	3.3757	2.7148

$$\mu\,(D) \quad x \quad y \quad z \quad \text{Total}$$
$$-0.004 \quad -0.036 \quad 0.032 \quad 0.04$$

5 THERMODYNAMIC PARAMETERS OF PHOSPHORIC ANHYDRIDE

The energy of one exocyclic phosphoryl bond is considered to be one-quarter of the total of the process [56] (Scheme 6a) which involves the formation of four P=O bonds.

$$P_4O_6\,(gas) + 4\,O\,(gas) \longrightarrow P_4O_{10}\,(gas) + 2336\,kJ/mol \tag{6a}$$

Some thermodynamic parameters of phosphoric anhydride that were determined experimentally [56–58] and calculated (MNDO/H [38], PM 3 [Authors] are presented in Table 2.3.

The experimental data for the vapour pressures of various modifications of phosphoric anhydride are presented in the publication [59].

It is noteworthy that the vapour pressure data determined for various modifications of phosphoric anhydride depend significantly upon the methods of preparation of the given forms as well as the instruments used. In the text book [59] the relation $P=f\,(T)$ is presented in the form of the following graph (Figure 2.1).

Table 2.3 Thermodynamic parameters of phosphoric anhydride [38]

T, K	$(U_t - U_0)$ (kJ/mol)	C_p J/mol K	C_{vib} J/mol K
298	28.8	190.3	157.1
300	29.2	191.2	157.9
400	49.4	227.5	194.2
500	72.7	253.1	219.9
600	98.1	271.3	238.0
700	125.1	284.3	251.0
800	153.2	293.8	260.5
900	182.1	300.8	267.6
1000	211.6	306.2	272.9
1100	241.6	310.3	277.1
1200	272.0	313.6	280.3

ΔH_f° (kJ/mol)	−2906.2	−2902.1 [38]	−1807.1 [38]	−2978.8 [Auth]
E_{total} (eV)			−3848.2	−3435.1
I_e (eV)	13.6±0.5		13.0	11.3

The thermodynamic parameters of phosphoric anhydride not listed in Table 2.3 are presented below [60].

Heat of sublimation	22.7 kcal/mol
Heat of vaporization	16.2 kcal/mol
Heat of fusion	6.5 kcal/mol
Standard free energy of formation	ΔG^0_{298} −644.8 kcal/mol

6 SPECTROSCOPIC METHODS OF STUDY

The spectroscopic information on phosphoric anhydride is rather limited predominantly because of its extremely low solubility [61]. Some spectroscopic data are reviewed above in the context of the molecular structure of phosphoric anhydride and in Chapter 3 devoted to its chemical properties.

Gerdering in collaboration with de Decker [46] and Van Brederode [62] delivered the first study of phosphorus oxides by means of Raman spectroscopy in 1946. The Raman spectra contained 12 fundamental lines two of which were overlapping. On the basis of analysis of the lines, the authors [46, 62] attributed the P_4O_{10} molecule to the tetrahedral point group T_d with the phosphorus atoms occupying the apexes of the regular tetrahedron.

In 1951 Daasch and Smith published a paper [63] describing the IR spectra of a large number of organo-phosphorus compounds (including P_4O_{10}). The

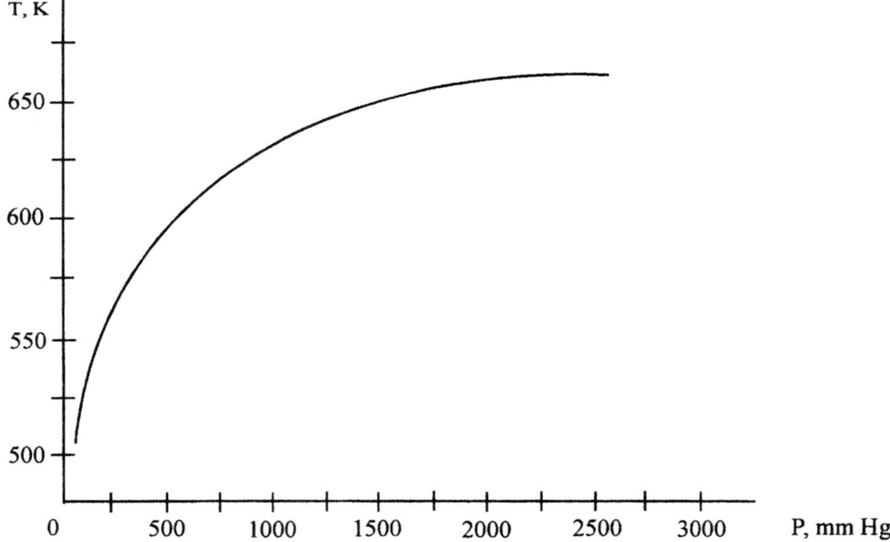

Figure 2.1 The vapour pressure of rhombohedral form of phosphoric anhydride at various temperatures

resolution was poor and no comments were made regarding the phosphoric anhydride spectrum.

Six years later Sidorov and Sobolev [47] studied the IR spectra of phosphoric anhydride in depth. The significance of this study is based not only upon the detailed description of the experimental data but also upon a very accurate comparative analysis of their own results and the Raman spectra data [46]. This greatly improved the authors' analysis of the molecular structure of phosphoric anhydride.

The Raman spectrum [46] of phosphoric anhydride contained the following set of lines, v, cm^{-1}: 257 (6), 278 (3), 329 (1), 424 (6), 559 (10), 650 (2), 721 (6), 952 (0, br), 1033(0, br), 1386 (4), 1417 (10), 1678 (1, br).

The IR spectrum [47] of phosphoric anhydride contained the sequence of bands, v, cm^{-1}: 573 (str), 714 (str), 764 (str), 832 (w), 843 (w), 1015 (v str), 1140 (w), 1150 (w), 1200 (w), 1231 (v w), 1295 (w), 1390 (str), 1590 (v w), 1648 (w), 1713 (w).

The more basic lines in the IR spectrum means that it is impossible for P_4O_{10} to belong to either of the point groups C_2, C_s, C_{3v}, $C_{\infty v}$ because the structures with such symmetry should have an equal number of bands in both types of spectra. Another characteristic feature of the spectra is that there are three similar lines, v (cm^{-1}): 573, 1015, 1390, which denies the possibility for P_4O_{10} to belong to the point groups of the types C_i, C_{2h}, $D_{2h} \equiv V_h$, S_6, C_{4h}, C_{6h}, $D_{3d} \equiv S_{6v}$,

$D_{4d} \equiv S_{8v}$, D_{4h}, D_{5h}, D_{6h}, $D_{\infty h}$, O_h because there should be no frequencies active simultaneously in Raman and IR spectra for these point groups. Another important band of the IR spectrum ($764\,\mathrm{cm}^{-1}$) is not present in the Raman spectrum which shortens the list of possible point groups for the structure of phosphoric anhydride molecules. Eventually on the basis of the spectroscopic data the formula of the compound was determined to be P_4O_{10}. It is noteworthy that the IR spectrum [47] contained two bands of somewhat high frequencies attributed to the P=O bond that supports its double character with some triple bond character that correlates well with the electron diffraction data presented above (Section 3).

On the basis of Raman spectroscopy and polarization data it was established that the 'cage' structure characteristic for the crystalline form of P_4O_{10} is retained in the gas phase [64] as well.

Later studies of phosphoric anhydride by means of IR spectroscopy were further developed [47]. Robinson [65] worked out the correlation between bond order (bond length) and force constants in P_4O_{10} and some other compounds of phosphorus, silicon, and chlorine. This correlation was based upon the author's assumption that only $3d_{x^2-y^2}$ and $3d_{z^2}$ orbitals of phosphorus atoms participate in the formation of the double bond with oxygen's unshared pair of electrons. The P=O bond order was determined to be 2.18.

Modern instruments and methods have made it possible to achieve very interesting and sophisticated results concerning phosphorus oxides. Withnall and Andrews [66] mixed phosphine with oxygen in argon and trapped the products in solid argon. The IR spectra of the solid argon matrix showed the formation of the molecules PO, HPO, PO_2, and the unknown PO_3, HOPO, P_2O_5 (Scheme 7), H_2PO, and HPOH. The molecule P_2O_5 was formed in the reaction of PO_2 with PO_3 [66] (Scheme 7).

$$P_n + O_2 \longrightarrow PO_2 + PO_3 \longrightarrow$$

(Argon matrix)

(Argon matrix) (7)

A similar method was applied by Lohr [37], who studied the structure of phosphorus oxides with the common formula P_2O_x ($x=1$–5) and characterized them by means of IR spectroscopy and *ab initio* electronic structure calculations (see also Section 4). The IR spectrum of P_2O_5 contains bands at $1473.2\,\mathrm{cm}^{-1}$ (P=O stretch, symm), $1158\,\mathrm{cm}^{-1}$ (P=O stretch, antisymm), $735.1\,\mathrm{cm}^{-1}$ (P–O–P antisymm). The following dimensions were calculated for the stereo structure of P_2O_5: P=O 1.435 (1.428) Å, P–O 1.568 (1.592) Å, P–O–P

131.6°, O=P=O 136.3°, O=P–O 111.7°. Though the structures of P_4O_{10} and P_2O_5 differ significantly, the considerable interest in the characteristic parameters of the latter arises from the fact that according to Lohr [37] the phosphoric anhydride molecule is a dimer (9) of P_2O_5.

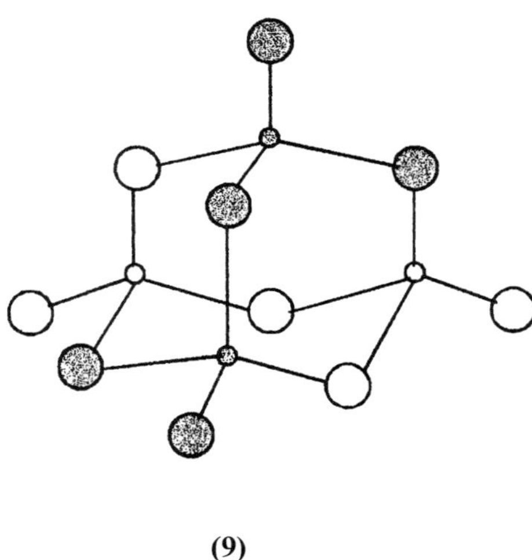

(9)

Further information on the Raman and IR studies of phosphoric anhydride are also presented in the publications [7, 64, 67]. One of the recent works devoted to IR spectroscopy of phosphoric anhydride was presented by Slivko and Krivoviazov [38]. The authors calculated the normal vibrations of P_4O_{10}, analysed the data in the valence-force field approximation and compared these with the experimental data [7, 64, 68]. The structure of phosphoric anhydride molecules was attributed [38] to the point symmetry group T_d which is characterized by 15 normal vibrations of the symmetry types $Å=3A_1+3E+3F_1+6F_2$. The six latter frequencies are present in the IR spectrum and the frequencies of all other types are registered in the Raman spectrum. The vibration spectrum contains the bands assigned to P=O bond: 1440 (symm), 1408 (antisymm), and $717\,cm^{-1}$. The 'cage' P–O bonds are characterized by the frequencies 1026, 1002, 824, 767, and $553\,cm^{-1}$. The authors [38] made an interesting comparative analysis of the latter group of bands with similar ones of the P_4O_6 IR spectrum (953, 817, 691, 641, and $613\,cm^{-1}$) [69]. The band shift in the region from 953 to $1026\,cm^{-1}$ in P_4O_{10} is explained by the interaction between P–O and P=O bonds with the input of the P=O bond in the frequency being $c.\,27\%$. The participation of the whole fragment O=P–O in the vibrations of the P–O bond results in the corresponding frequencies increasing from 691 and $641\,cm^{-1}$ in P_4O_6 up to 824 and $767\,cm^{-1}$, respectively in P_4O_{10}.

The papers [7, 38] present the IR spectra of the isotope-substituted analogues of phosphoric anhydride $P_4O_9{}^{18}O$, $P_4O_8{}^{18}O_2$, $P_4O_7{}^{18}O_3$, and $P_4O_6{}^{18}O_4$. The $P_4O_6{}^{18}O_4$ species in solid argon were identified by the ^{18}O band $1365.6\,cm^{-1}$ which was shifted for $42.3\,cm^{-1}$ in comparison with unlabelled P_4O_{10} [7]. Under similar conditions the authors characterized P_4O_{10} by the bands 1408, 1026, 768 and $577\,cm^{-1}$ [7].

CHAPTER 3

Reactivity of Phosphoric Anhydride

Organophosphorus compounds have been synthesized for many decades using a limited range of phosphorylating agents such as PCl_3, PH_3, $POCl_3$, $PSCl_3$, P_4O_{10}, P_4S_{10} and to a less extent iodo-, fluoro- and other inorganic phosphorus-containing compounds and elemental phosphorus [1, 2, 70–72]. The other principal publications are reviewed below in the relevant sections.

The outstanding feature of phosphoric anhydride is its ability to phosphorylate organic compounds, mostly nucleophilic reagents, with the direct formation of acidic phosphates. This feature is undisputed across all the publications devoted to P_4O_{10} that we have reviewed.

1 REACTIONS OF PHOSPHORIC ANHYDRIDE WITH OXYGEN-CONTAINING COMPOUNDS

1.1 HYDRATION OF PHOSPHORIC ANHYDRIDE

The early experimental data concerning the hydration of phosphoric anhydride were rather unreliable. Since the nineteenth century there was a belief that under hydrolysis of phosphoric anhydride diphosphoric acid was formed directly. In 1940 Glioxelli published two papers [73, 74] in which he described the unknown acids derived from phosphoric anhydride [73] and the electroconductivity of aqueous solutions prepared from different crystalline types of phosphoric anhydride [74].

The first realistic scheme (Scheme 8) for the hydration of phosphoric anhydride was based not upon experimental evidence but was worked out theoretically by Khodakov in 1944 [22]. A stepwise process was described with the formation of intermediate tetrametaphosphoric and isotetrametaphosphoric acids, though the latter intermediate is presented as less favourable. The scheme suggested [22] was proved experimentally six years later by Rodionova and

Khodakov [75]. The sublimed phosphoric anhydride was treated with water at 17–19°C either in acidic (H_2O) or in basic (NaOH) media. After a very quick process of phosphoric anhydride 'loosening', two molecules of water were added and tetrametaphosphoric acid $H_4P_4O_{12}$ (Scheme 8) was formed. The subsequent reaction broke down the cyclic structure of the latter to give tetraphosphoric acid $H_6P_4O_{13}$ that under further hydrolysis produced orthophosphoric acid and triphosphoric acid $H_5P_3O_{10}$ and so on via pyrophosphoric acid down to orthophosphoric acid entirely (Scheme 8). The formation of tetra, tri, and diphosphoric acids was determined by titration which was a complicated experimental procedure as described in reference [75].

Hydration of P_4O_{10} in a basic medium proceeded faster than in an acidic medium [22, 75]. It was assumed that pyrophosphoric acid was not formed earlier than 1.5–2 h. Its presence in the reaction mixture was determined by titration only after 3 h [75], which indicates a step-by-step reaction of the P_4O_{10} tricyclic structure with water.

$$(8)$$

$$\xrightarrow{\ \ H_2O\ \ } HO-\underset{\underset{OH}{|}}{\overset{\overset{O}{\|}}{P}}-O-\underset{\underset{OH}{|}}{\overset{\overset{O}{\|}}{P}}-O-\underset{\underset{OH}{|}}{\overset{\overset{O}{\|}}{P}}-O-\underset{\underset{OH}{|}}{\overset{\overset{O}{\|}}{P}}-OH \xrightarrow[-H_3PO_4]{H_2O}$$

$$H_6P_4O_{13}$$

$$\longrightarrow HO-\underset{\underset{OH}{|}}{\overset{\overset{O}{\|}}{P}}-O-\underset{\underset{OH}{|}}{\overset{\overset{O}{\|}}{P}}-O-\underset{\underset{OH}{|}}{\overset{\overset{O}{\|}}{P}}-OH \xrightarrow[-H_3PO_4]{H_2O}$$

$$H_5P_3O_{10}$$

$$\longrightarrow HO-\underset{\underset{OH}{|}}{\overset{\overset{O}{\|}}{P}}-O-\underset{\underset{OH}{|}}{\overset{\overset{O}{\|}}{P}}-OH \xrightarrow{H_2O} 2H_3PO_4$$

$$H_4P_2O_7$$

The formation of a tetrametaphosphoric intermediate in this process was claimed also by the authors of publication [76]. The suggestion that hydrolysis of phosphoric anhydride proceeds via the formation of isotetrametaphosphoric acid (**9a**) was rejected in paper [75].

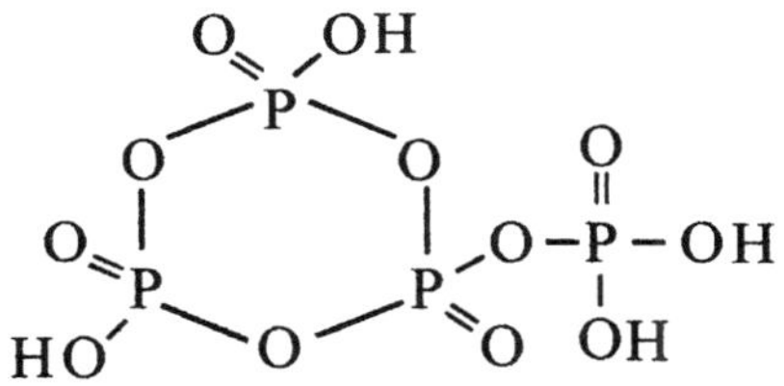

(9a)

The poor exchange of scientific information between Russian and Western specialists in the 1940–50s probably limited access to Khodakov's papers [22, 75] for the authors of publication [77], which may explain why the formation of tetrametaphosphoric acid in the hydrolysis experiments with phosphoric anhydride at low temperature came as a surprise to Bell and co-workers [77]. The authors [77] also did not consider the formation of tetrametaphosphoric acid with the iso-structure.

Contrary to the earlier data, Thilo and Wieker, and Ratz [78, 79] in the early 1950s established, along with tetrametaphosphoric acid [80], the presence of iso-tetrametaphosphoric acid in the reaction mixture of phosphoric anhydride with water and proved this fact by isolating the relatively stable trimetaphosphoric acid H$_3$P$_3$O$_9$ which had to be the product of one-step hydration of isotetrametaphosphoric acid (Scheme 9).

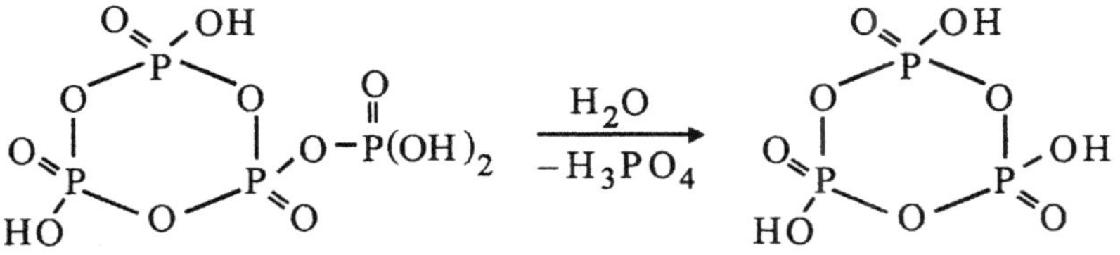

$$\text{(9)}$$

Thilo and Wieker [78] also described the hydrolytic reactivity of various crystalline forms of P$_4$O$_{10}$.

There is relatively little spectroscopic information on the hydrolysis intermediates, which probably reflects the limited number of studies of the hydrolysis of phosphoric anhydride and the difficulty in isolating many of the intermediates. The indirect study of the products of hydration of P$_4$O$_{10}$ was carried out by Van Wazer and co-workers [8] in a reverse process of their reconstruction of phosphoric anhydrides from phosphoric acid. The bicyclic dianion (**10**) which could be presented as the product of one molecule of water added to P_4O_{10} was synthesized in the reaction of dry orthophosphoric acid with carbodiimide in tetramethylurea (see also Chapter 1). Its ^{31}P NMR spectroscopic data were determined and assignments confirmed by computer simulation (δ −28.9, −36.5 ppm, J^2_{PP} 30.4 Hz).

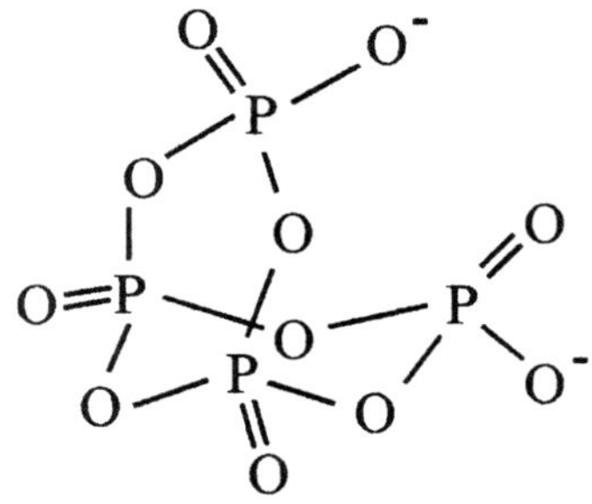

(**10**)

The observed and calculated spectra correlated well. Another very important result of the same research was the formation of a disubstituted anion of isotetrametaphosphoric acid. Again the computer simulated ^{31}P NMR spectrum (δ −13.5, −24.1, −33.5 ppm) was in good agreement with the experimental spectrum.

Hydrolysis of these intermediates proceeded via triphosphoric and diphosphoric acids [8]. Among several possible pathways for the hydration of isotetrametaphosphate (**9a**), the authors [8] preferred two paths; the most favourable mechanisms included the cleavage of an endocyclic P–O–P bond and the other involved scission of the exocyclic P–O–P bond.

Analysis of meta-, pyro- and orthophosphates or their mixtures was carried out in some cases [82 and references therein] by investigating the hydration of phosphoric anhydride in aqueous sulphuric or nitric acids. Such studies did not target the elucidation of the process pathway but concentrated on distinguishing the above mentioned phosphoric acids by various titrimetric methods.

Chromatographic analysis of the products of phosphoric anhydride hydration carried out by Henry, Nickless and Pollard [83] under neutral conditions at 0–5°C revealed a high ratio of species built up by six phosphorus atoms. Moreover phosphoric anhydride vapour heated at 200–300°C produced polymeric phosphate (possibly octapolyphosphate) chains. Another remarkable result of the study [83] revealed a pronounced stability of trimetaphosphate under prolonged (90 min and longer) heating at 100°C.

Shima and co-authors [84] investigated phosphoric anhydride hydration in water, and various organic solvents such as anhydrous acetone, ethanol, diethyl ether, and mixed systems. The data accumulated for the direct hydration by water at 0°C supported the earlier results published by Thilo and co-workers [78, 79] and Bell and co-authors [77]. The reaction in pure acetone gave surprisingly mainly tetrametaphosphoric and high polyphosphoric acids. The solvolysis by neat acetone remains unclear because the nature of the interaction between P$_4$O$_{10}$ and acetone is not specified. The hydrolysis of phosphoric anhydride with aqueous solutions of ethanol was accompanied by partial esterification of P$_4$O$_{10}$ and produced mostly pyrophosphoric acid and its diethyl ester, along with dialkyl tetrametaphosphoric acid. An increase in the water content of the solution directed the process towards the formation of tetrametaphosphoric acid at the expense of pyrophosphate [84], the reason for which is not obvious. The process of pyrophosphate formation is presented as the synchronous split of two P–O–P bonds in tetrametaphosphoric acid leading to symmetrically disubstituted and nonsubstituted pyrophosphoric acids (Scheme 10).

The possibility of monosubstituted pyrophosphoric acid formation by splitting of the disubstituted tetrametaphosphoric acid was not considered by the authors [84].

Dillon and Waddington [85] studied hydration of phosphoric anhydride under specific conditions of sulphuric acid, chlorosulphuric acid and oleum media (in the range of SO$_3$ concentrations 0–65%). An interesting feature of phosphoric anhydride solutions in 100% H$_2$SO$_4$ stems from the exceptionally strong dehydrating activity of P$_4$O$_{10}$, which leads one to consider such systems as solutions of phosphoric anhydride and the initial products of its hydration in dilute oleum (concentration of SO$_3$ less than 20%). An increase of oleum concentration suppressed the extent of phosphoric anhydride hydrolysis as indicated by phosphorus-31 NMR. The low concentration solutions of P$_4$O$_{10}$ in 20% oleum were characterized by two groups of signals in the ranges from -3.6 to 9.9 ppm and from -16.8 to -14.5 ppm. The high field signals, according to the authors [85], indicated the presence of the protonated form of pyrophosphoric acid. The amount of the higher molecular weight phosphoric acids increased with increase of SO$_3$ concentration in oleum.

Phosphorus-31 NMR spectra patterns of the solutions of phosphoric anhydride in chlorosulphuric acid were more complicated. Besides two groups of signals at 0.1 ppm and -14.3 ppm, a signal appeared at -34.3 ppm which was present only in freshly prepared solutions. This high field signal was attributed to the branched phosphorus atoms formed as a result of the first step of phosphoric anhydride hydrolysis (see Structure **10**). The data presented in the publication [85] strongly support the prevailing formation of trimetaphosphate cyclic structures as a result of the addition of the third molecule of water to P$_4$O$_{10}$. In other words, phosphoric anhydride hydration pathways in sulphuric acid proceed predominantly via isotetrametaphosphoric acid (Scheme 11) with the following cleavage of the exocyclic P–O–P bond.

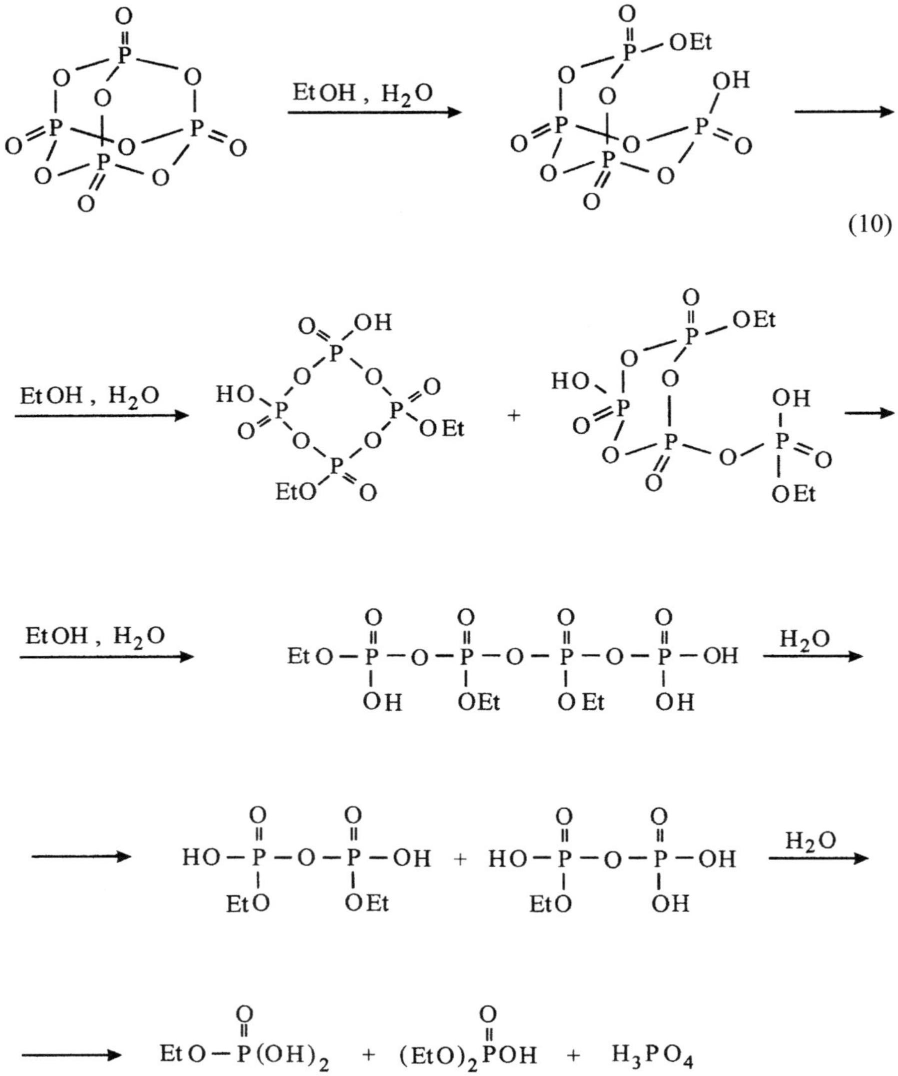

$$(10)$$

In aqueous solutions of phosphoric anhydride at 0°C the ratio of isotetrametaphosphoric acid to cyclotetrametaphosphoric acid was determined to be about 0.80 [86].

The chemical shift of the exocyclic dihydroxyphosphoryl group in ^{31}P NMR spectra of isotetrametaphosphate is not specified by the authors [86]. The overall scheme of phosphoric anhydride hydration in sulphuric acid matches well in many respects with its direct reaction with water, alcohols (Section 1.2), amines (Section 3.2) and other nucleophilic reagents.

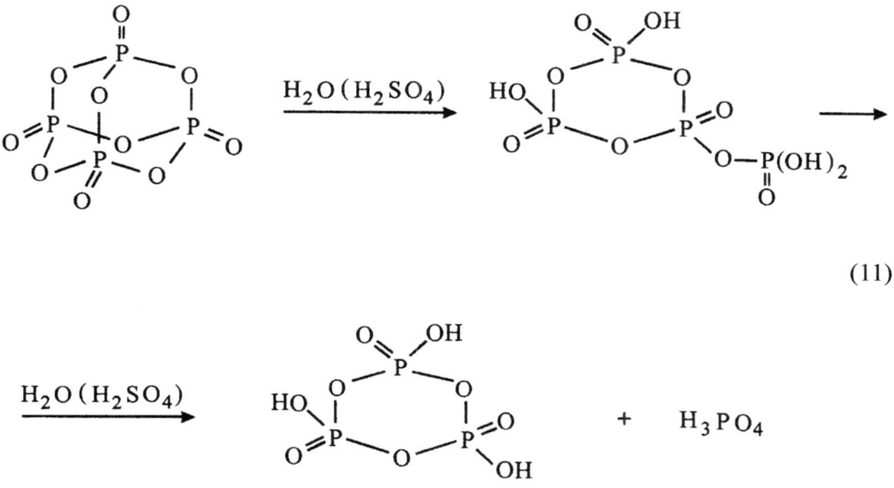

$$(11)$$

The equilibrium sets of compounds formed from various phosphates (KH$_2$PO$_4$, NaO-P(O)(OH)-O-P(O)(OH)Na, Na$_5$P$_3$O$_{10}$) in 100% sulphuric acid and oleum were similar to those formed from phosphoric anhydride in the same media [86].

The controlled reaction of P$_4$O$_{10}$ with water followed by neutralization with NaOH has been used for the commercial manufacture of sodium tetrametaphosphate Na$_4$P$_4$O$_{12}$, which crystallizes from solution after the addition of sodium chloride [77, 87].

In conclusion the mainstream stages of the process of phosphoric anhydride hydration can probably be presented by the overall Scheme 12.

Unlike phosphoric anhydride, the corresponding thioanhydride is rather stable under the action of the atmospheric moisture and a faint smell of hydrogen sulphide indicates the slow process of hydrolysis. But under acidic conditions it undergoes fast hydration to form phosphoric acid and hydrogen sulphide (Scheme 13). This is contrary to phosphoric anhydride, which is more stable in acidic media than in neutral or basic media.

The rate of P$_4$S$_{10}$ hydrolysis is promoted significantly in the presence of the catalytic quantities of mercury iodide [88].

In basic media some salts of thiophosphoric acid were formed as the byproducts (Scheme 14a). The latter compounds are characterized by thiophosphoryl–phosphoryl equilibria of the type shown in Scheme 14b.

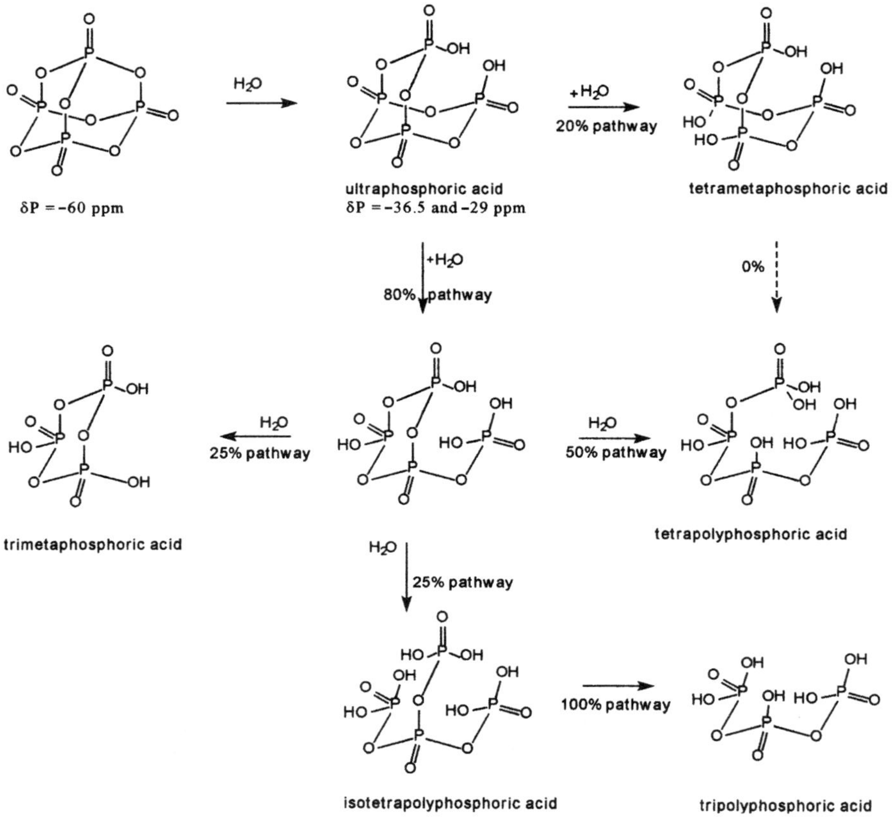

$\delta P = -60$ ppm

ultraphosphoric acid
$\delta P = -36.5$ and -29 ppm

$+H_2O$
20% pathway

tetrametaphosphoric acid

$+H_2O$
80% pathway

0%

H_2O
25% pathway

trimetaphosphoric acid

H_2O
50% pathway

tetrapolyphosphoric acid

H_2O
25% pathway

isotetrapolyphosphoric acid

100% pathway

tripolyphosphoric acid

$$\tag{12}$$

$$P_4S_{10} + 16\,H_2O \longrightarrow 4\,H_3PO_4 + 10\,H_2S \tag{13}$$

$$P_4S_{10} + 12\,NaOH \longrightarrow 4\,Na_3PO_2S_2 + 2\,H_2S + 2\,H_2O \tag{14a}$$

$$\left[\begin{array}{l} NaO \\ NaO \\ NaS \end{array}\!\!\diagdown\!\!\diagup P{=}S \rightleftharpoons \begin{array}{l} NaO \\ NaS \\ NaS \end{array}\!\!\diagdown\!\!\diagup P{=}O \right] \tag{14b}$$

1.2 REACTIONS WITH ALCOHOLS AND PHENOLS

In 1910 Dupuis [88] published a paper devoted to the reactions of phosphoric anhydride with substituted phenols. The process between P$_4$O$_{10}$ and p-methoxyphenol resulted in the formation of a mixture of disubstituted and monosubstituted phosphates (Scheme 15). The yield of the former product predominated. The exothermic process (causing heating up to 100°C) was carried out without a solvent [88]. The products were isolated in the form of the corresponding potassium or sodium salts.

$$\text{MeO}-\langle\bigcirc\rangle-\text{OH} \xrightarrow[\text{H}_2\text{O}]{\text{P}_4\text{O}_{10}} \left(\text{MeO}-\langle\bigcirc\rangle-\text{O}\right)_2 \overset{\text{O}}{\overset{\|}{\text{P}}}\text{OH} + \text{MeO}-\langle\bigcirc\rangle-\text{O}\overset{\text{O}}{\overset{\|}{\text{P}}}\text{(OH)}_2$$

$$(15)$$

Later Grun and Limpacher [90–93] made a significant input to the development of phosphoric anhydride chemistry. They introduced P$_4$O$_{10}$ to reactions with weakly reactive disubstituted glycerols that resulted in the formation of the corresponding phosphorylated derivatives (Scheme 16) including those of lecithins and kephalins.

$$(16)$$

$$\text{Alk} = \text{C}_{17}\text{H}_{35}$$

The reactive hydroxyl group in glycerols could be either terminal [91] or bonded to the middle carbon atom [92].

Such an interesting specific approach to phosphorylated glycerols was used in the synthesis of surfactants: it involved the partial basic hydrolysis at C2 of triolein from vegetable oil with consequent reaction with phosphoric anhydride [94] (Scheme 17). (See also Chapter 6.)

$$\begin{array}{c} H_2C-O-\overset{\overset{\textstyle O}{\|}}{C}\,Alk \\ | \\ H\overset{|}{C}-O-H \\ | \\ H_2C-O-\underset{\underset{\textstyle O}{\|}}{C}\,Alk \end{array} \quad \xrightarrow[\text{2. } NH_4OH]{\text{1. } P_4O_{10}} \quad \begin{array}{c} H_2C-O-\overset{\overset{\textstyle O}{\|}}{C}\,Alk \\ | \\ H\overset{|}{C}-O-PO_3(NH_4)_2 \\ | \\ H_2C-O-\underset{\underset{\textstyle O}{\|}}{C}\,Alk \end{array} \qquad (17)$$

$$Alk = C_{17}H_{35}$$

The probable pathways of the reactions (Scheme 18) that produced derivatives of metaphosphoric acid, monosubstituted, disubstituted and mixed esters of phosphoric acid are discussed [93].

$$(18)$$

Such an approach to mixed esters of phosphoric acid is of particular importance in bioorganic chemistry, however the stepwise reactions of phosphoric anhydride with two or more alcohols was not further developed.

A convenient path to monoalkyl and dialkyl phosphoric acids was proposed by Plimmer and Burch [95], the method involved the reaction of phosphoric anhydride with ethanol, chloroethanol, cetyl alcohols, cholesterol, and other alcohols (Scheme 19). Similar results were achieved [95] in the reactions of the same alcohols with either orthophosphoric acid (Scheme 20) or phosphoryl chloride (Scheme 21). Which method is the most advantageous remains unclear though the latter reagent is likely to be the least favourable.

$$R\,OH \;+\; P_4O_{10}$$

$$\Big\downarrow H_2O \quad (19)$$

$$POCl_3 + ROH \;\xrightarrow{(21)}\; R\underset{O}{\overset{O}{\underset{\|}{P}}}(OH)_2 + (RO)_2\underset{O}{\overset{O}{\underset{\|}{P}}}OH \;\xleftarrow{(20)}\; ROH + H_3PO_4$$

The same target was achieved much later by Cherbuliez and co-authors [96] in the reaction of phosphoric anhydride with a wide range of alcohols of low reactivity. The process was carried out in a reacting alcohol media and supported by pyridine. The individual products were isolated by converting the resulting phosphates into corresponding barium salts and their crystalline hydrates (**11–13**). It is noteworthy that this method was applied to an efficient synthesis of unsaturated esters with ethylenic or acetylenic groups. No spectroscopic characteristics were presented for the compounds (**11–13**) [96].

$$CH_3CH{=}CH{-}CH_2O\underset{O}{\overset{}{\underset{\|}{P}}}O_2^{-2}\,Ba^{+2}\cdot H_2O \qquad\qquad CH_2{=}CH{-}CH_2O\underset{O}{\overset{}{\underset{\|}{P}}}O_2^{-2}\,Ba^{+2}\cdot 2H_2O$$

$$\textbf{(11)} \qquad\qquad\qquad\qquad \textbf{(12)}$$

$$HOCH_2C{\equiv}C{-}CH_2O\underset{O}{\overset{}{\underset{\|}{P}}}O_2^{-2}\,Ba^{+2}\cdot 2H_2O$$

$$\textbf{(13)}$$

A mixture of phosphates possessing the distinctive features of an active surfactant was synthesized in the reaction of phosphoric anhydride with hydroxyalkyl ester produced by esterification of stearic acid by 2,2-dimethylpropanediol [97] (Scheme 22).

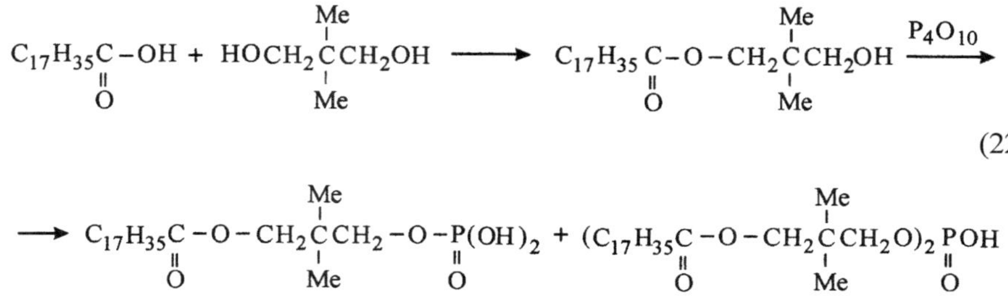

$$C_{17}H_{35}\overset{\text{O}}{\underset{\|}{C}}-OH + HOCH_2\overset{Me}{\underset{Me}{C}}CH_2OH \longrightarrow C_{17}H_{35}\overset{\text{O}}{\underset{\|}{C}}-O-CH_2\overset{Me}{\underset{Me}{C}}CH_2OH \xrightarrow{P_4O_{10}}$$

$$(22)$$

$$\longrightarrow C_{17}H_{35}\overset{O}{\underset{\|}{C}}-O-CH_2\overset{Me}{\underset{Me}{C}}CH_2-O-\overset{O}{\underset{\|}{P}}(OH)_2 + (C_{17}H_{35}\overset{O}{\underset{\|}{C}}-O-CH_2\overset{Me}{\underset{Me}{C}}CH_2O)_2\overset{O}{\underset{\|}{P}}OH$$

According to Cherbuliez [98, 99], Ivanov [100], Petrov and Treschalina [101] the reaction of phosphoric anhydride with alcohols is a stepwise process and the P–O–P bonds become more stable with decrease in the number of the –O–P groups linked to a phosphorus atom. It is presumed that the process starts with the protonation of hemipolar PO_4 fragments. The tertiary amines inhibit either step of interaction between phosphoric anhydride and alcohols [100]. A method for a wide variety of monoalkyl, monoaryl, dialkyl and diaryl phosphoric acids based on the reaction of P_4O_{10} with alcohols was developed by Okamoto [102].

According to the classical view on the nucleophilic reactions the interaction between phosphoric anhydride and alcohols must start with the protonation of the phosphoryl oxygen followed by cleavage of a POP bond (Scheme 23).

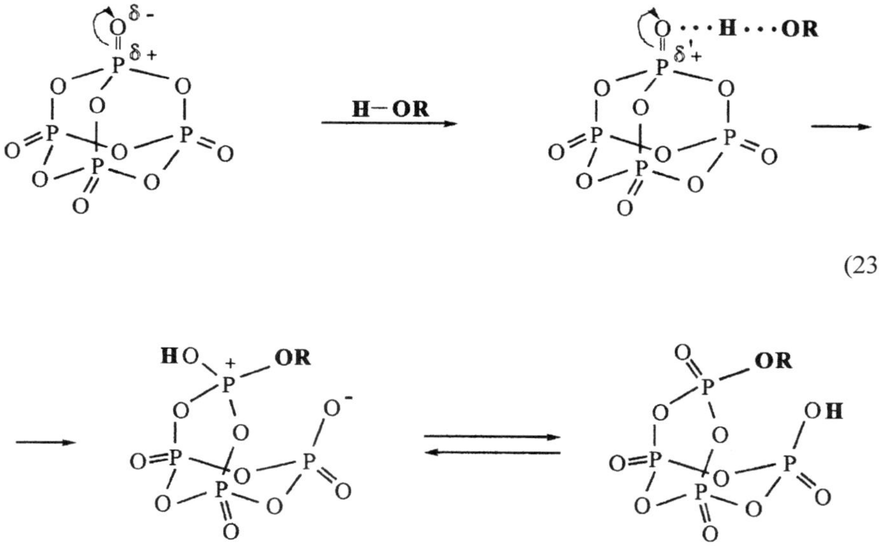

$$(23)$$

This mechanism contradicts the effect of the very high stability of the P=O bond due to its exceptionally short dimension which was emphasized by Hampson and Stosick [24] and other scientists.

For this reason Kosolapoff [71] expressed the primary step of interaction between phosphoric anhydride and alcohols as orientation of the hydroxy group along the P–O bond of phosphoric anhydride (Scheme 24).

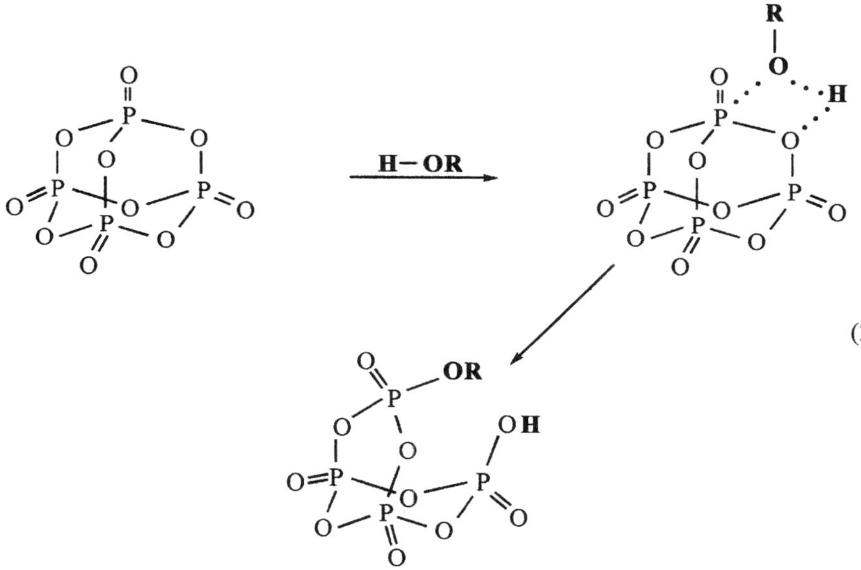

$$(24)$$

While such an interpretation of the process is presented in many publications including text-books [103], it seems likely to the authors that several pathways may be involved and that most will involve more than one molecule of alcohol.

Peridoxal phosphate was synthesized in the reaction of phosphoric anhydride with a corresponding Schiff base (**14**) [104]. The intermediate polyphosphate was hydrolysed by hydrochloric acid under mild conditions.

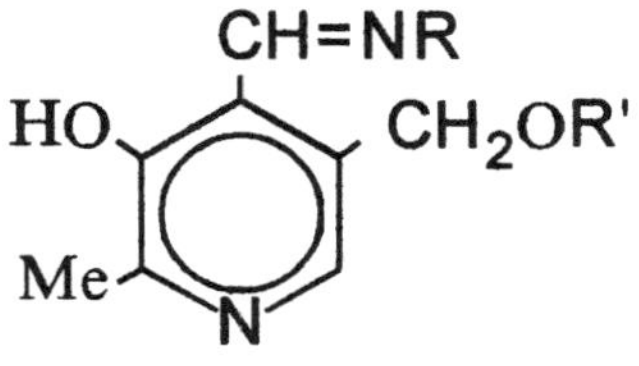

(14)

The naturally occurring meta-substituted phenol, cardanol, reacted readily with phosphoric anhydride in an inert medium at 50–55°C to give a symmetrically disubstituted pyrophosphate with the yield 80% (Scheme 25) [105]. Subsequent acidic hydrolysis of pyrophosphate resulted in the formation of the corresponding monoester of phosphoric acid (90%), which was polymerized into monocardanyl phosphoric acid-formaldehyde resin possessing a very high thermal stability [105]. The IR and ^{1}H NMR but no ^{31}P NMR spectra are presented in the publication [105].

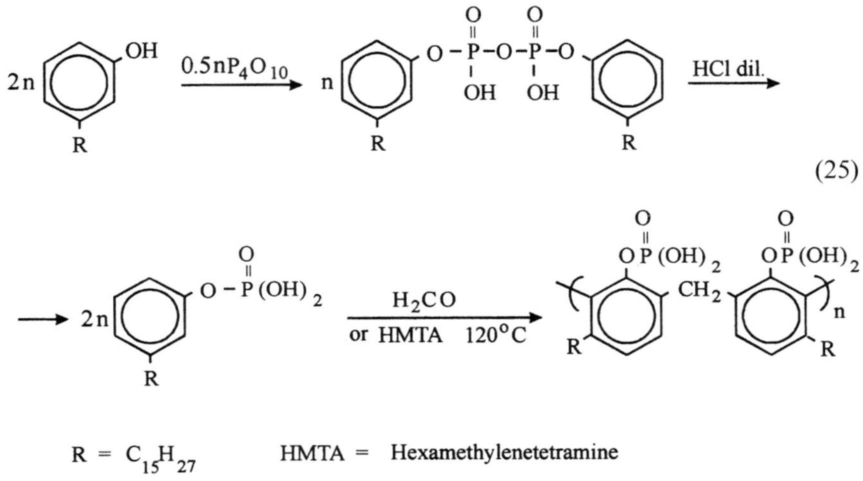

$$2n \ \text{ArOH} \xrightarrow{0.5nP_4O_{10}} n \ \text{Ar–O–P(O)(OH)–O–P(O)(OH)–O–Ar} \xrightarrow{\text{HCl dil.}}$$

$$\tag{25}$$

$$\longrightarrow 2n \ \text{Ar–O–P(O)(OH)}_2 \xrightarrow[\text{or HMTA} \quad 120^\circ C]{H_2CO} \left[\text{Ar(OP(O)(OH)}_2)\text{–CH}_2\text{–Ar(OP(O)(OH)}_2) \right]_n$$

$$R = C_{15}H_{27} \qquad \text{HMTA} = \text{Hexamethylenetetramine}$$

Mixed esters of phosphoric acid may also be synthesized. Thus phosphoric anhydride and a mixture of alcohols gave the diester shown [106] (Scheme 26).

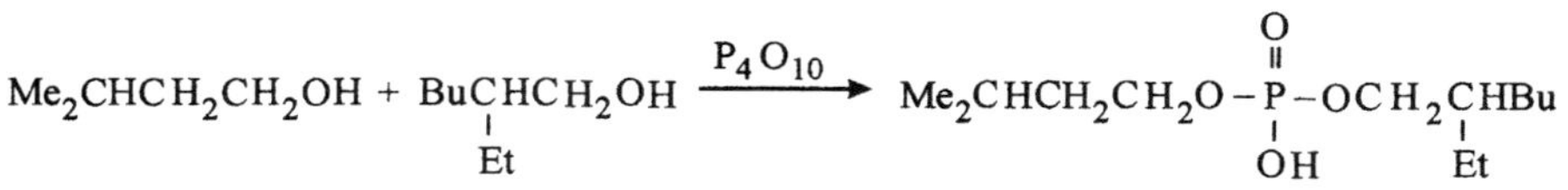

$$\text{Me}_2\text{CHCH}_2\text{CH}_2\text{OH} + \text{BuCH(Et)CH}_2\text{OH} \xrightarrow{P_4O_{10}} \text{Me}_2\text{CHCH}_2\text{CH}_2\text{O–P(O)(OH)–OCH}_2\text{CH(Et)Bu}$$

$$\tag{26}$$

The reaction took place in hexane which acted not only as the medium for the reaction but also as a selective solvent for the dialkyl phosphate formed. A mixture of both alcohols was added to phosphoric anhydride at temperatures below 40°C. A similar reaction occurred in light mineral oil medium [106] (Section 6.4).

Adler and Woodstock [107] claimed the formation of alkyl polyphosphates from phosphoric anhydride and either 2-ethylhexanol or *n*-octanol (Scheme 27).

$$5\,Alk-OH + 1.5\,P_4O_{10} \longrightarrow Alk_5-P_6O_{20}H_5 \qquad (27)$$

$$Alk=CH_2\underset{\underset{\displaystyle C_2H_5}{|}}{C}H(CH_2)_3CH_3,\ C_8H_{17}$$

The latest detailed study of the mechanism of phosphoric anhydride reaction with alcohols was by Elias, Azzouz, and Rodehuser [108] in 1993. The authors described an interaction between P_4O_{10} and 2-ethylhexanol without a solvent (Scheme 28) as in the earlier works [108a]. The process was monitored step by step by means of ^{31}P NMR spectrometry. Eventually the formation of orthophosphoric acid, mono(2-ethylhexyl) phosphoric acid (δ_{31P} 0.9–1.7 ppm), di(2-ethylhexyl) phosphoric acid (δ_{31P} from −0.5 to 0.5 ppm), and tri(2-ethylhexyl) phosphoric acid (δ_{31P} from −1 to −2 ppm) was established. A triplet at −27 ppm was attributed to the central phosphorus atoms (P^c) of symmetrically substituted triphosphoric acid (**15**). An important ^{31}P NMR signal with the shift −13 ppm belonged to symmetrically disubstituted pyrophosphoric acid; other signals in the region −11 to −15 ppm were assigned to pyrophosphoric acid, its unsymmetrically substituted esters and the terminal phosphorus atoms of triphosphoric acid. The Handbook [109] gives a more general account of the trends in ^{31}P NMR chemical shifts, including the influence of pH.

$$P_4O_{10}+Alk\,OH \longrightarrow Alk\,O\underset{\underset{\displaystyle O}{\|}}{P}(OH)_2 + (Alk\,O)_2\underset{\underset{\displaystyle O}{\|}}{P}OH + (Alk\,O)_3P{=}O \qquad (28)$$

$$Alk=C_8H_{17}$$

$$HO(Alk\,O)\underset{\underset{\displaystyle O}{\|}}{P}-O-\underset{\underset{\displaystyle O}{\|}}{P^c}(OR)-O-\underset{\underset{\displaystyle O}{\|}}{P}(O\,Alk)OH$$

$$R = H,\ C_8H_{17}$$

$$(\mathbf{15})$$

The process was studied at temperatures of 21, 35, 78 and 95°C [108]. After 22 h of heating within this range of temperatures an insignificant proportion of triphosphoric acid was present at 21°C. Pyrophosphates were detected at 21 and 35°C. At 21°C the ratio of products changed insignificantly during the whole period of observation. The ratio of di-(2-ethylhexyl) phosphoric acid grew steadily up to 78°C and then decreased at 95°C [108]. Traces of tri-(2-ethylhexyl) phosphoric acid were present only at 78°C. At all temperatures the proportion of mono-(2-ethylhexyl) phosphoric acid was the highest. This publication is of particular practical importance because mixtures of alkyl phosphoric and phosphoric acids prepared in the reaction of 2-ethylhexanol with phosphoric anhydride are produced industrially without a solvent on a broad scale and are widely used as active surfactants. However, the actual content of such mixtures was never scrutinized in detail.

The complicated process of starch phosphorylation was simplified by preliminary treatment of starch with dry sodium carbonate or dry potassium hydroxide at 30–100°C followed by reaction with phosphoric anhydride [110]. The second step of the process took 30–120 min at 80–100°C.

Phosphorylated cellulose and its derivatives are known to possess a number of useful properties. The direct phosphorylation of cellulose by phosphoric anhydride took place in an organic solvent that dissolved cellulose and was inert towards phosphoric anhydride [111]. The solvents used were ethylene chloride, tetrachloroethane, chloroform and tricresyl phosphate. (Incidentally the former two compounds are known to be moderately reactive towards phosphoric anhydride under certain conditions.) The process lasted for 10–24 h to give phosphorylated cellulose with rather low content of phosphorus (about 1–2%).

A more advanced method for phosphorylating cellulose and its derivatives was worked out by Clermont [112, 113]. Compounds phosphorylated included methyl cellulose, commercial water-soluble hydroxymethyl cellulose, hydroxyethyl cellulose, cellulose nitrate (content of nitrogen 10.0%), partially acylated cellulose (content of acetyl groups 25%), cotton batting and filter paper. Prior to phosphorylation phosphoric anhydride reacted with dimethylformamide to give a stable complex. The following interaction of cellulose with that complex took place under mild conditions (temperature range 20–100°C); even at the room temperature the process was complete within 1–15 min. The products were water-insoluble compounds with phosphorus content up to 10% (corresponding to phosphorylated hydroxyethyl cellulose). The lower rate of phosphorus (1–2%) was typical for partially acylated cellulose, cotton batting and filter paper.

The phosphorylated cellulose derivatives had pronounced ion-exchange activity when the phosphorus content rose above $c.\,4\%$, i.e. it increased in direct proportion to the phosphorus content. For example, phosphorylated methyl cellulose (content of P 5.6%) possessed the ion-exchange capacity 1.15 milliequivalent per 1 g of dry adsorbent [113].

Though the author of the patent [113] gave examples only of small-scale reactions, the whole procedure including following make-up was very simple and moreover clean from the environmental point of view. Thus the filtrate after isolation of the solid product could be reused in the next reactions until the complete exhaustion of P_4O_{10}–DMF complex. Dimethyl formamide was easily recycled by distillation. The phosphorylation reaction can be adapted for a continuous process [113]. No information concerning the structure of products was given [113].

A wide range of polyfluoro phosphoric acids was synthesized by Benning [114] in a reaction of phosphoric anhydride with the corresponding linear polyfluoro alcohols. The latter compounds were prepared by polymerization of tetrafluoroethylene catalysed by either peroxy or azo compounds at 50–350°C accompanied by substitution of the terminal fluorine by the hydroxyl group [115] (Scheme 29).

Polyfluoroalkanols reacted with phosphoric anhydride in the molar ratio 1:4 and the following hydration of the intermediate bis(polyfluoroalkyl) pyrophosphoric acid gave the corresponding polyfluoromonoalkyl phosphoric acids (Scheme 29). These products were isolated either neat or in the form of their ammonium salts. Pyrolysis was used to produce bis(polyfluoroalkyl) phosphoric acids of the corresponding bis(polyfluoroalkyl) pyrophosphoric acids at very low pressure [115] (Scheme 29). In all experiments the process of pyrolysis was accompanied by the formation of tris(polyfluoroalkyl) phosphoric acids and the initial polyfluoro alcohols.

$$0.5\,n\,HCF_2-CF_2H \xrightarrow{\text{MeOH, Cat.}} H(CF_2)_n CH_2OH \xrightarrow{P_4O_{10}}$$

$$\xrightarrow{} H(CF_2)_n CH_2O\overset{O}{\underset{OH}{\overset{\|}{P}}}-O-\overset{O}{\underset{OH}{\overset{\|}{P}}}OCH_2(CF_2)_n H \xrightarrow{H_2O} 2\,H(CF_2)_n CH_2O\overset{O}{\overset{\|}{P}}(OH)_2$$

$$\Big\downarrow 200\text{-}300°C$$

$$[H(CF_2)_n CH_2O]_2 \overset{O}{\overset{\|}{P}}-OH \quad + \quad [H(CF_2)_n CH_2O]_3 P{=}O \quad + \quad H(CF_2)_n CH_2OH$$

(29)

$$n = 2, 4, 6, 8, 10$$

The influence of the length of alkyl and alkoxyalkyl radicals upon the content of the mainstream products (monoalkylphosphates, dialkylphosphates, phosphoric acid) in reactions with phosphoric anhydride was considered by O'Lenick and Parkinson [122]. From an investigation of a series of alkoxydecyl alcohols, higher yields of monoalkyl phosphoric acids and orthophosphoric acid were obtained at the expense of dialkyl phosphoric acids in the order $CH_3(CH_2)_9OH$, $CH_3(CH_2)_9O(CH_2CH_2O)_2H$, $CH_3(CH_2)_9O(CH_2CH_2O)_6H$, $CH_3(CH_2)_9O(CH_2CH_2O)_8H$. High yields of monoalkylphosphates were also obtained for another set of analogues of the type $C_{13}H_{27}O(CH_2CH_2O)_n$ H ($n=0, 3, 7, 9, 12$). This effect is explained entirely in terms of steric hindrance. The ratio of products formed was determined by acid–base titration. Unfortunately the publication [122] does not contain the experimental section for the phosphorylation of the alcohols.

Nusslein [123] described the synthesis of a wide variety of phosphate esters from phosphoric anhydride and glycol, polyglycols and fatty alcohols (Scheme 31).

$$AlkOH \xrightarrow{P_4O_{10}} (AlkO)_3 P=O$$

$$HOC_nH_{2n}OH \xrightarrow{P_4O_{10}} (HOC_nH_{2n}O)_3 P=O \xrightarrow{Alk^1COOH} \tag{31}$$

$$\longrightarrow (Alk^1\overset{\text{O}}{\underset{\|}{C}}OC_nH_{2n}O)_3 P=O$$

It is noteworthy that other workers [123] have cited trialkyl phosphates as the prevailing products of the process. Moreover [123] the formation of cyclic diphosphates (Compound **16**) and mixed diphosphates (Compound **17**) was also reported. Some of the structures claimed [123] need additional scrutiny by means of more modern methods.

$$O=P(OC_nH_{2n}O)_3 P=O$$

$$O=P\overset{\textstyle OAlk}{\underset{\textstyle OC_nH_{2n}O\overset{\|}{\underset{O}{P}}(OH)_2}{\big\langle}} OC_nH_{2n}OAlk^1$$

$$(16) \qquad\qquad\qquad\qquad\qquad (17)$$

Woodstock [124] described a different approach to the synthesis of potential surfactants. The method for mixed alkyl phosphates consisted of prior phosphorylation of alcohols with phosphoric anhydride followed by condensation of monoalkyl phosphoric acid with ethylene oxide (Scheme 32).

$$C_{12}H_{25}OH \xrightarrow{P_4O_{10}} C_{12}H_{25}O\overset{O}{\overset{\|}{P}}(OH)_2 + (C_{12}H_{25}O)_2\overset{O}{\overset{\|}{P}}-OH$$

$$\tag{32}$$

$$C_{12}H_{25}O\underset{\underset{O}{\|}}{P}(OH)_2 + 2n\ CH_2-CH_2 \longrightarrow C_{12}H_{25}O\underset{\underset{O}{\|}}{P}[(OCH_2CH_2)_n OH]_2$$

Phosphorylation of alcohols by mixtures of phosphoric anhydride with a third reagent is recommended in numerous publications. In most cases these third reagents are orthophosphoric acid with or without water, phosphorus oxychloride, alkyl polyphosphate ester, methanesulphonic acid, sulphuric acid, dimethyl sulfoxide, tertiary amines and metals oxides. It should be noted that such reactions do not involve the neat P_4O_{10} as a phosphorylating agent but the corresponding products of its reaction with the additional reagent, e.g. to give hydration (Section 1.1) and/or products of its interaction with H_3PO_4, $MeSO_3H$, $POCl_3$ (Section 4) that are poly, tri and/or diphosphoric acids and other corresponding derivatives. Nevertheless we find it relevant to present a brief review of such reactions. The wide and efficient applications of P_4O_{10} in phosphorylating mixtures, particularly with phosphoric acid, water or polyphosphoric acid, is characteristic of bioorganic chemistry (Chapter 5).

The growing industrial demand for a rich variety of monoalkyl, dialkyl and trialkyl phosphoric acids that are applicable as surfactants, fire retardants, lubricants and anticorrosive agents inspired several very fruitful studies illustrated in the literature and patents that are of considerable academic and industrial interest.

One rare in-depth comparative study of the phosphorylation of alcohols by phosphoric anhydride and the mixtures phosphoric anhydride–phosphoric acid, phosphoric anhydride–polyphosphoric acid was presented by a group of Japanese scientists [125]. The detailed scrutinizing of the reaction of phosphoric anhydride with alkanols (C_8–C_{12}) in the presence of water exhibited its nonequilibrium character. The authors singled out the main factors influencing the process. Among these factors, the ratio nucleophilic reagents (alcohol+water):electrophilic reagent (phosphoric anhydride) was determined to be the most influential, and was the factor that determined the overall result of the process. Secondly the authors [125] considered the effect of the sequence of the reaction steps upon the nature of the products formed i.e.

a) hydrolysis of phosphoric anhydride followed by phosphorylation or b) phosphorylation of an alcohol by neat phosphoric anhydride followed by hydrolysis. Traditionally the industrial approach involves carrying out the reactions without a solvent. The processes were thoroughly monitored by means of ^{31}P NMR technique which allowed all the components of the products to be identified and also determined their ratio. The reaction pathways for the processes 'a' and 'b' differed significantly. In the course of the reaction, the quantity of phosphoric acid remained unchanged though the breakdown of the pyrophosphate bonds P–O–P must result in the formation of phosphoric acid. This was explained [125] by simultaneous direct esterification of phosphoric acid by alcohol. It was concluded that there was insignificant temperature influence upon the process in the range 50–95°C.

In the case with the molar ratio nucleophile:electrophile equal to 3:1 the quantity of monoalkyl phosphoric acid was inversely proportional to the ratio AlkOH:H$_2$O within its range 1.5–2.5. According to the charts presented in the paper [125] the procedure with the forthcoming phosphorylation of alcohol by neat phosphoric anhydride, followed by hydrolysis, was preferable for the formation of monoalkyl phosphoric acid. The yield of the latter product grew in time as determined by ^{31}P NMR.

Unlike the earlier findings of Cherbuliez [126] the authors [125] phosphorylated *n*-dodecanol by phosphoric acid (100%) under azeotropical distillation of the water formed from the reaction mixture with xylene. Targeting the industrial application of this reaction the process was carried out at the reduced pressure which allowed the maximum yield (80%) of monododecyl phosphoric acid (Scheme 33) to be achieved. Unfortunately the product was rather unstable under the reaction conditions due to its pyrolysis which made the method unsuitable as an industrial process.

$$\text{n–C}_{12}\text{H}_{25}\text{–OH} \quad \xrightarrow{\text{H}_3\text{PO}_4 \text{ or PPA}} \quad \text{n–C}_{12}\text{H}_{25}\text{–O–}\underset{\overset{\|}{\text{O}}}{\text{P}}(\text{OH})_2 \qquad (33)$$

Unlike phosphoric acid an excess of pyrophosphoric acid phosphorylated *n*-dodecyl alcohol under mild conditions to produce the corresponding monoalkyl phosphoric acid in nearly quantitative yield accompanied by traces of dialkyl phosphoric acid. The optimum molar ratio P$_4$O$_{10}$:C$_{12}$H$_{25}$OH was found to be between 4:1 and 5:1. By the end of the process the residual polyphosphoric acid had to be removed.

A remarkable point of the data presented in the publication [125] is the formation of both symmetrically and asymmetrically substituted pyro-phosphates (Figure 3.1) in the reaction mixture. These are of particular importance for considering the mechanism of the process. Contrary to [8], the

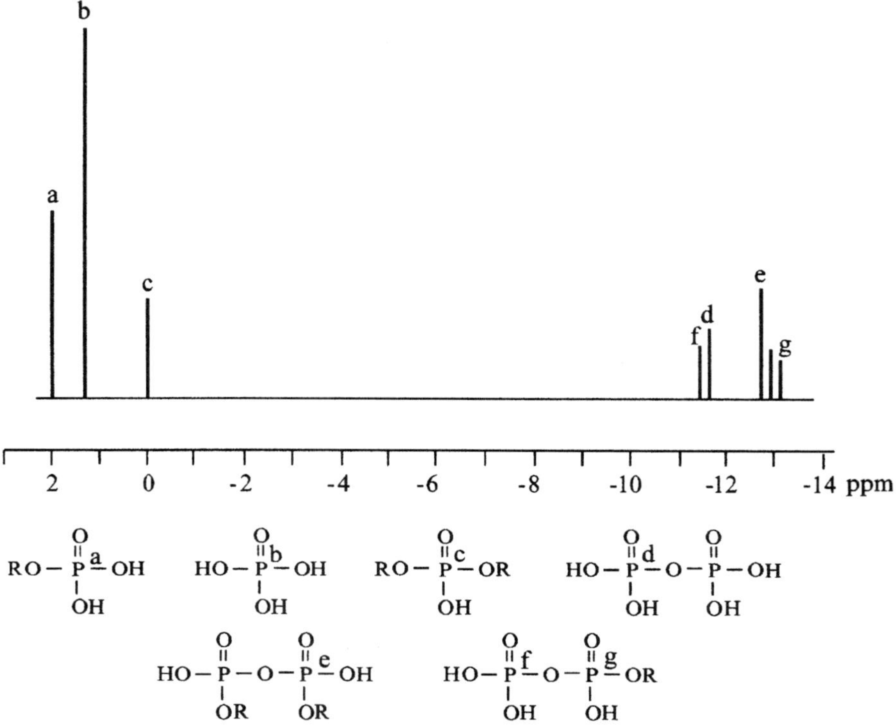

Figure 3.1 Phosphorus-31 NMR spectra of the reaction mixture P_4O_{10}—AlkOH—H_2O [125]

authors [125] found the triphosphates and pyrophosphates to be unstable, particularly in their acidic forms.

The influence of pH upon the rate of 5% solutions of hydrolysis of monoalkyl phosphoric acids was also studied [125]. The maximum rate was achieved at pH 6.4 (90°C). High molecular weight alcohols have a tendency to produce a mixture of phosphates which is not necessarily a disadvantage for industrial applications.

Monoalkyl phosphate was also reported by a group of Japanese scientists [127] to be the dominating product of the interaction between N,N-disubstituted ethanolamine and phosphoric anhydride/phosphoric acid. 2-(N-dodecyl-N-methyl) ethanolamine was synthesized in the reaction of dodecylbromide with N-methylethanolamine in ethanol under basic conditions (Scheme 34). Dialkylamine reacted in tetrahydrofuran under mild conditions (room temperature, 20 min) with 85% H_3PO_4 (molar ratio amine:H_3PO_4=1:1) followed by addition of phosphoric anhydride (molar ratio H_3PO_4:P_4O_{10}=1:0.7) [127] (Scheme 34).

$$CH_3(CH_2)_{11}Br \ + \ HN\Big\langle{\!\!\!\!\!^{Me} \atop \!\!\!\!\!_{CH_2CH_2OH}} \xrightarrow{\ Na_2CO_3(EtOH)\ } CH_3(CH_2)_{11}-N\Big\langle{\!\!\!\!\!^{Me} \atop \!\!\!\!\!_{CH_2CH_2OH}} \longrightarrow$$

$$(34)$$

$$\xrightarrow[\ 2.\, P_4O_{10}\ ]{1.\, H_3PO_4\ (THF)} CH_3(CH_2)_{11}-N\Big\langle{\!\!\!\!\!^{Me} \atop \!\!\!\!\!_{CH_2CH_2O-\underset{O}{\overset{OH}{P}}-O-P(OH)_2}} \xrightarrow[\ 2.\, NaOH\ ]{1.\, H_2O}$$

$$\longrightarrow CH_3(CH_2)_{11}-N\Big\langle{\!\!\!\!\!^{Me} \atop \!\!\!\!\!_{CH_2CH_2O-\underset{O}{\overset{OH}{P}}-ONa}} \ + \ NaH_2PO_4$$

After refluxing for 8 h the reaction mixture was hydrolysed with water and treated with sodium hydroxide to bind the phosphoric acid [127] (Scheme 34). The product was isolated with the yield 65% and purified by column chromatography.

The potentiometric titration of 2-(N-dodecyl-N-methylamino) ethyl phosphoric acid registered three differential maxima that corresponded to the values pK_1 4.7–4.8, pK_2 8.8–8.9, pK_3 9.4–9.5. The pH value of its aqueous solution differed insignificantly from the isoelectric point. Analysis of these data lead the authors [127] to the conclusion that N,N-dialkylaminoethyl phosphoric acid could exist in four ionic forms (Scheme 35) which are dependent on the acidity of its aqueous solutions.

$$\text{Alk}-\overset{+}{\underset{H}{N}}\Big\langle{\!\!\!\!\!^{Me} \atop \!\!\!\!\!_{CH_2CH_2O-\underset{O}{\overset{OH}{P}}-OH}} \rightleftharpoons \text{Alk}-\overset{+}{\underset{H}{N}}\Big\langle{\!\!\!\!\!^{Me} \atop \!\!\!\!\!_{CH_2CH_2O-\underset{O}{\overset{OH}{P}}-O^-}} \rightleftharpoons$$

$$\text{pH} < 4.7 \qquad\qquad\qquad \text{pH}\ \ 4.7-4.8 \qquad\qquad (35)$$

$$\rightleftharpoons \text{Alk}-N\Big\langle{\!\!\!\!\!^{Me} \atop \!\!\!\!\!_{CH_2CH_2O-\underset{O}{\overset{OH}{P}}-O^-\ H^+}} \rightleftharpoons \text{Alk}-N\Big\langle{\!\!\!\!\!^{Me} \atop \!\!\!\!\!_{CH_2CH_2O-\underset{O}{\overset{O^-}{P}}-O^-}}\ \ 2\,H^+$$

$$\text{pH} > 8.9 \qquad\qquad\qquad\qquad \text{pH} > 8.9$$

Czech chemists [128] studied the kinetics of the interaction between diacyl glycerols, phosphoric anhydride and metaphosphoric acid (Scheme 36). The

When the molar ratio of the reagents $P_4O_{10}:H(CF_2)_nCH_2OH$ were $1:6$, a $50:50$ mixture of mono(polyfluoroalkyl) phosphoric acid and bis(polyfluoroalkyl) phosphoric acid was produced. The individual products and their mixtures were analysed by potentiometric titration. The yields of the individual products were reported to be up to 79% [115]. The halogenoalkyl phosphoric acids were determined to be active surfactants [116, 117].

The main problem for the industrial production of individual monoalkyl (aryl), dialkyl (aryl) and trialkyl (aryl) phosphoric acids or mixtures of these compounds in a desired ratio, and minimum content of residual polyphosphates and phosphoric acid, is that reactions of phosphoric anhydride with alcohols produce multicomponent mixtures. For instance, Hasigawa [118] attempted to produce polymerizable monomers containing a phosphoryl group by the phosphorylation of an alcohol using phosphoric anhydride. According to the patent application [118], copolymerization of a mixture of dialkyl and trialkyl phosphates with other monomers resulted in an unwanted process of gelation. Phosphates which polymerized efficiently with vinyl or acrylyl monomer phosphates were produced in the reaction with the limited molar ratio of the initial reagents $(P_4O_{10}:ROH=1.0:1.8$ or more) and under rigid temperature control $(20-100°C)$. Exceeding the top temperature caused polymerization of unsaturated phosphoric esters and the ratio of $1.8\,mol$ of alcohol per $1.0\,mol$ of phosphoric anhydride helped to suppress the formation of phosphoric acid which otherwise was difficult to remove.

These results agreed with Milwidsky and Gabriel [119], i.e. an excess of phosphoric anhydride in a reaction with alcohols is an unfavourable factor for the technological processes because of the formation of residual polyphosphoric acids. A mild boiling of such mixtures for the hydrolysis of polyphosphoric and pyrophosphoric chains was proposed.

In many cases industrial surfactants are produced by reacting alcohols with ethylene oxide followed by phosphorylation by phosphoric anhydride in accordance with Scheme 30.

$$\text{Alk}\,CH_2-OH \ + \ \underset{\diagdown \, O \, \diagup}{CH_2-CH_2} \ \longrightarrow \ \text{Alk}\,CH_2-O-CH_2\,CH_2-OH \ \xrightarrow{P_4O_{10}}$$

$$(30)$$

$$\longrightarrow \ \text{Alk}\,CH_2-O-CH_2\,CH_2-O-\underset{O}{\overset{\parallel}{P}}(OH)_2 \ + \ (\text{Alk}\,CH_2-O-CH_2\,CH_2O)_2\underset{O}{\overset{\parallel}{P}}OH$$

A Japanese patent describes the synthesis of monoalkyl and dialkyl phosphates by the reaction of long-chain alcohols (C8–C36) with phosphoric anhydride [120, 121].

process, which was studied in an adiabatic reactor, was characterized by a reaction order higher than 1. The enthalpy value was determined to be 50–60 kJ/mol, and the reaction halftime at 80–100°C was several minutes. The increase of the initial quantity of phosphoric anhydride per hydroxyl groups shortened the reaction halftime radically.

$$
\begin{array}{ccc}
\begin{array}{l} CH_2OAc \\ | \\ CHOAc \\ | \\ CH_2OH \end{array}
& \xrightarrow{\ P_4O_{10}\ } &
\begin{array}{l} CH_2OAc \\ | \\ CHOAc \quad O^- \\ | \qquad\quad | \\ CH_2O-\overset{\displaystyle O}{\underset{||}{P}}-O^- \end{array}
\quad + \quad
\begin{array}{l} (CH_2OAc \\ | \\ CHOAc \\ | \\ CH_2O)_2\,\overset{\displaystyle O}{\underset{||}{P}}-O^- \end{array}
\end{array}
\qquad (36)
$$

Unlike reactions of phosphoric anhydride with alcohols or phenols the phosphorylating mixture P_4O_{10}–H_3PO_4 presented a higher possibility to direct the process towards full esters of phosphoric acid. Substituted phenols reacted with phosphoric anhydride and orthophosphoric acid to give triphenyl phosphate [129] (Scheme 37), which was not the case in the reactions studied by Dupuis [88].

$$
3\,Ar\text{–}OH \quad \xrightarrow{\ P_4O_{10}/H_3PO_4\ } \quad (ArO)_3P{=}O \qquad (37)
$$

Phosphorylation of adenosine by phosphoric anhydride in a water medium with the formation of 2′- and 3′-adenosinemonophosphate [130] is presented in Chapter 5.

Urea has also been used as a component of phosphorylating mixtures based on phosphoric anhydride. Thus starch was phosphorylated by P_4O_{10}–H_2NCONH_2 (1.1:2) in high yield [131]. The other method of starch phosphorylation by phosphoric anhydride was presented earlier in this section.

An intriguing approach to alkylaryl phosphoric acids proposed by Otto [132] consisted of simultaneous alkylation of the aromatic ring by an olefin and phosphorylation of phenol by phosphoric anhydride. P_4O_{10} was added to a preheated (40–50°C) mixture of phenol and an olefin over a period of several hours maintaining the temperature of the exothermic process between 90 and 110°C (Scheme 38).

Though the most favourable ratio of an alkene to phosphoric anhydride was 6:1 other proportions resulted in the predominant formation of either product of the mixture. Higher ratios of alkene could give the corresponding dialkylaryl phosphoric acids and even trialkylaryl phosphoric acids [132]. The range of olefins introduced in the reaction was very broad and included the compounds with the terminal double bond such as 2,4,4-trimethyl-1-pentene, 2-ethyl-1-hexene, 2,3-dimethyl-1-hexene, α-methyl styrene, 2-methyl-1-penta-decene, tri- and with tetrasubstituted double bonds such as 2-methyl-2-butene,

2-methyl-2-pentene, 2,5-dimethyl-2-hexene, 2,4,4-trimethyl-2-pentene, 3,4-dimethyl-2-butene, 2,3-dimethyl-2-pentene, 3,4-dimethyl-3-hexene and 2,3,4-trimethyl-2-pentene. Nevertheless there were some types of ethylenes that could not be used by this method. Generally the olefins suitable for the reaction possessed at least one disubstituted sp^2 hybridized carbon atom. In some cases mixtures of alkenes were used, for example a mixture of 2,4,4-trimethyl-1-pentene and 2,4,4-trimethyl-2-pentene was found to be a very active regent. Otto [132] did not study the actual mechanism of the reaction but presented a probable scheme for the process according to which the very first step was phosphorylation of phenol by phosphoric anhydride with the prevailing formation of monoaryl and diaryl phosphoric acids (Step 1) (Scheme 39) that acted as the catalyst of phenol alkylation (Step 2) (Scheme 39); this was similar to the well-known alkylation catalysts such as sulphuric acid, boron trifluoride, hydrogen fluoride, zinc chloride, aluminium chloride and some others. The third step was phosphorylation of an alkylphenol by either phosphoric anhydride, its residual fragments, or the phosphorylated phenol (Scheme 39). The final step resulted in the regeneration of some of the initial unsubstituted phenol.

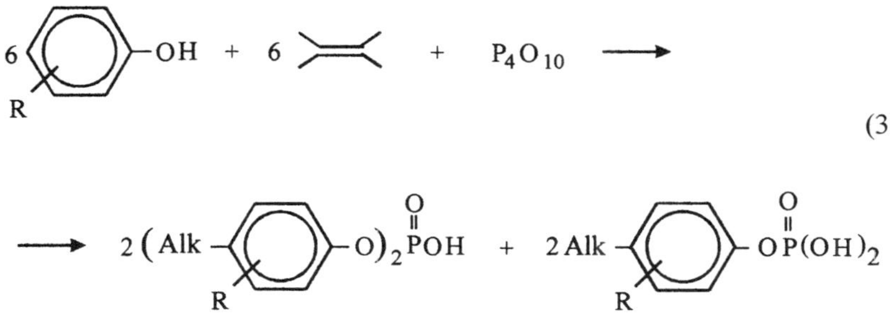

$$(38)$$

Hence, the specific feature of this reaction was the double action of phosphoric anhydride which included phosphorylation of the hydroxyl group of both unsubstituted and alkylated phenols.

So, phosphoric anhydride itself or in combination with another supporting reagent is shown to be a good phosphorylating agent towards alcohols and phenols of low reactivity. It can be anticipated that the reactions of highly active alcohols with P_4O_{10} should not necessarily terminate with the formation of corresponding phosphates but proceed further involving for example the processes of dehydration (Sections 8.1).

Step 1

Step 2

(39)

Step 3

$R = Alk$

Polyphosphoric acid interaction with alcohols was scrutinized in a detailed experimental and statistical study performed by Clarke and Lyons [133] which resulted in the determining of some unorthodox data. It is interesting that the process under consideration was presented by the authors as 'the alcoholysis of polyphosphoric acid' but not phosphorylation of alcohols by polyphosphoric acid emphasizing probably the mainstream direction of their investigations. The reactions of alcohols with polyphosphoric acid containing 84.3% of P_4O_{10} which corresponded to 117% H_3PO_4 were performed in a similar manner to the study [108] presented in Section 1.2. The reactions were performed without a solvent excluding the influence of the third component and with strict control of water content. The reactions had moderate exothermic effect not higher than 10–30°C and their completion was achieved after heating at 50–100°C (Scheme 40) as determined by a potentiometric titration [108].

$$(HO)_2P-O-\left[P-O\right]\left[P(OH)_2\right]_n + m\,ROH \longrightarrow m\,RO-P(OH)_2 + (n-m)H_3PO_4$$

(40)

$$R = C_4H_9,\ C_5H_{11},\ C_8H_{17},\ C_{10}H_{21},\ C_{13}H_{27},\ C_{16}H_{33}$$

The processes were also followed by phosphorus-31 NMR. The terminal phosphorus atoms were registered by the signals at $-12.0\,$ppm and the middle P atoms at $-29.4\,$ppm. The separation of overlapping signals was achieved by making sodium salts of the corresponding alkyl phosphates which resulted in a high-field shielding effect [134].

Practically all the reactions studied [134] ended up with the formation of monoalkyl orthophosphoric acids and free orthophosphoric acid with an alcohol–PPA ratio 1:4. In the case of t-butyl alcohol intramolecular dehydration took place (Scheme 41).

$$CH_3-\underset{\underset{CH_3}{|}}{\overset{\overset{CH_3}{|}}{C}}-OH \xrightarrow[-H_2O]{PPA} CH_3-\underset{\underset{CH_3}{|}}{\overset{\overset{CH_2}{\|}}{C}}$$

(41)

Intermolecular dehydration of alkanols with the formation of corresponding dialkyl ethers was not reported by the authors [134]. When the initial reaction mixture contained one mole of an alcohol per each P–O–P bond in polyphosphoric acid, diesters, $[(PO)_2PO_2H]$, (up to 14%) were formed.

Esterification of phosphoric acid with the formation of pyrophosphoric acid was not ruled out [134], but was said to be negligible. For pyrophosphoric acid the formation of dialkyl phosphates in the reaction with alcohols by breaking down the P–O–P bond was considered to be improbable and for the intermediate triphosphoric, tetraphosphoric and polyphosphoric acids a one-by-one split of the terminal P–O–P bonds with the formation of monoalkyl phosphoric acid and the one unit shorter polyphosphoric chain is proposed. In such a case this is the important distinguishing point between phosphoric anhydride and polyphosphoric acid. Three bridging P–O–P bonds in P_4O_{10} that build up its tricyclic structure direct the process towards dialkyl phosphates along with monoalkyl phosphoric acids and the accumulation of only a small quantity of orthophosphoric acid.

On the basis of statistical analysis [134] it was calculated that not more than 35% of dialkyl phosphates could be formed in a reaction of polyphosphoric acid with alcohols, but in earlier experiments [135] the yield of dialkyl phosphates had been determined to be only 1.5%. So, the general conclusion of Clarke and Lyons [135] was the random cleavage of the polyphosphate chain in polyphosphoric acid leads to monoalkyl phosphates. The presence of monohydroxyphosphoryl species among the products was attributed to the symmetrical dialkyl pyrophosphoric acids that were nonreactive under the experimental conditions.

Phosphoric thioanhydride reacted with alcohols with the formation of O,O-dialkyl esters of dithiophosphoric acids as the major products [136]. Malatesta and Laverone [137] were the first to prepare O,O-diethyl phosphoric ester by this reaction (Scheme 42).

$$P_4S_{10} + 8\,EtOH \longrightarrow 4\,(EtO)_2\,\underset{\overset{\|}{S}}{P}SH + 2\,H_2S \tag{42}$$

Other alkanethiols reacted with P_4S_{10} to give dithioesters of tetrathiophosphoric acid (Scheme 43) with rather high yields.

$$P_4S_{10} + 8\,RSH \longrightarrow 4\,(RS)_2\,\underset{\overset{\|}{S}}{P}SH + 2\,H_2S \tag{43}$$

1.3 REACTIONS WITH ETHERS

The first example of a reaction of phosphoric anhydride with ethers was described by Langheld [138, 139] at the beginning of this century. The viscous reaction mixture formed from phosphoric anhydride and diethyl ether was considered to be ethyl metaphosphate [138–142]. In the paper [141] this product was presented as an equilibrium with the asymmetrically substituted pyrophosphate (Scheme 44).

$$\underset{O}{\overset{O}{P}}\!\!-O-\underset{O}{P}(OEt)_2 \rightleftharpoons 2 \ \underset{O}{\overset{O}{P}}\!\!-OEt \tag{44}$$

The reaction of phosphoric anhydride with diethyl ether proceeds much smoother in chloroform [142].

Several decades later Thilo and co-workers [79, 143] discovered by chemical methods that the viscous product named 'Langheld-ester' was a multicomponent mixture (Scheme 45). The usefulness of Langheld-ester (polyphosphoric ester, PPE) nowadays is with respect to its high potential in bioorganic synthesis (Chapter 5).

According to authors of the publications [79, 143] the primary steps of the reaction in chloroform medium were directed towards formation of cyclic intermediates — substituted tetrametaphosphate and isotetrametaphosphate. Hydrolysis of Langheld-ester in boiling water gave a set of products: alkyl phosphoric acid, dialkyl phosphoric acid, 1,2-dialkyl pyrophosphoric acid, orthophosphoric acid and some other undetermined compounds. The ratio of the components was highly dependable upon the alkyl groups of the ethers. No trialkyl phosphate formation was reported in the publications [79, 143]. Slow hydrolysis of the mixture produced from the reaction of phosphoric anhydride and diisopropyl ether gave diisopropyl phosphoric acid (Scheme 46). This is attributed to the prevailing accumulation of isotetrametaphosphate in the reaction mixture.

The structure of diisopropylphosphoric acid was proved by its independent synthesis based on phosphorus trichloride [143] (Scheme 47). This route was much more complicated in comparison with that based on phosphoric anhydride.

The scheme for the reaction of P_4O_{10} with diethyl ether which was proposed by Ratz and Thilo [79] (Scheme 20) was not totally supported by the NMR data [144, 145]. Several publications reviewed below demonstrate the polemic character of the reaction mechanism.

^{31}P NMR studies of the crude reaction mixtures of phosphoric anhydride and ether [144, 145] indicate that the reaction does not proceed beyond the formation of isotetrametaphosphate, which is the main component of the reaction product. The spectra of such mixtures contained three signals with chemical shifts -16, -29, and -42 ppm that were assigned to terminal, middle and branched phosphorus atoms, respectively. Authors of the paper [145] published the time-dependent spectrum of the reaction mixture during hydrolysis. It should be emphasized that such spectra of ^{31}P NMR are characteristic also for reaction mixtures of phosphoric anhydride with other organic compounds (Sections 3.3.1–3.3.3).

$$\text{(45)}$$

$$RO-\underset{\underset{O}{\|}}{P}(OH)_2 \;+\; (RO)_2\underset{\underset{O}{\|}}{P}OH \;+$$

$$+\; RO-\underset{\underset{O}{\|}}{P}-O-\underset{\underset{O}{\|}}{P}-OR \;+\; H_3PO_4 \;+\; \ldots$$

with each P bearing an OH group.

R = Et, i-Pr

Pollmann and Schramm [146] improved the synthetic approach to Langheld ester and used another method for its analysis. Potentiometric titration of the resulting mixture in dimethyl formamide suggested that it consisted of iso-tetrametaphosphate (85%) and tetrametaphosphate (15%). These data strongly supported the results described earlier in the paper [79] but the authors [79, 147] did not consider the ^{31}P NMR spectra.

A strong band at $1300\,\text{cm}^{-1}$ in IR spectrum of this mixture (Langheld ester), which was recorded in a chloroform solution, was attributed to the cyclic metaphosphate structure [146, 148]. Under exposure to humid air this band vanished and another band characteristic for the open-chain polyphosphate esters appeared between 1250 and 1200 cm^{-1}. Potentiometric titration of the hydrolysis of isometatetraphosphate confirmed predominant hydration of the cyclic P–O–P bond rather than the exocyclic P–O–P bond [146].

$$P\,Cl_3 \; + \; 3\;i\text{-PrOH} \;\longrightarrow\; (i\text{-PrO})_3\,P \;\xrightarrow{\;HCl\;}\; (i\text{-Pr}\,O)_2\,\underset{\underset{O}{\|}}{P}H \;\xrightarrow{\;Cl_2\;}$$

$$\longrightarrow\; (i\text{-Pr}\,O)_2\,\underset{\underset{O}{\|}}{P}-Cl \qquad\qquad\qquad\qquad\qquad (46)$$

$$H_2O\,(C_5H_5N)$$

$$H_2O \;\longrightarrow\; (i\text{-PrO})_2\,\underset{\underset{O}{\|}}{P}-OH \qquad\qquad (47)$$

$$P_4O_{10} \; + \; i\text{-Pr}_2O \;\longrightarrow\; PPE$$

Synthetic approaches to hexalkyl tetraphosphate [149] and neutral esters of phosphoric acid [150, 151] from phosphoric anhydride and ethers has been patented.

A detailed study of the phosphoric anhydride reaction with diethyl ether was performed by Adler [149]. The most important outcome of this research was the determination of the optimum temperature range, which was found to be crucial for the formation of various individual products of this reaction. At temperatures below 90°C tetraethyl metaphosphate was formed (Scheme 48). This compound was naturally an intermediate on the reaction pathway directed towards the formation of hexaethyl tetraphosphate (Scheme 48). This step was carried out at strictly controlled temperatures between 90 and 110°C [149].

At higher temperatures (above 120°C) the reaction proceeded towards the pyrophosphate [149] (Scheme 48) which means that hexaethyl tetraphosphate reacted with the excess of diethyl ether by splitting the central P^2–O–P^3 bond. The small amount of orthophosphoric acid that could be formed in the process of disproportionation gave strong support to such an effect. The process was carried out in an autoclave with slow steady heating to favour the formation of tetraethyl metaphosphate followed by 7–8 h of heating at 90–110°C. The ratio of phosphoric anhydride–diethyl ether was not important, except that a small excess of the latter reagent had to be used [149]. Insignificant amounts of the esters of meta and orthophosphoric acid were present in the product (which contained the equivalent of 57–60% of P_2O_5). This data corresponded well to the general formula of hexaethyl tetraphosphate (56% P_2O_5) containing small impurity of tetraethyl metaphosphate. Such a procedure was used for making a large number of industrial size batches [149].

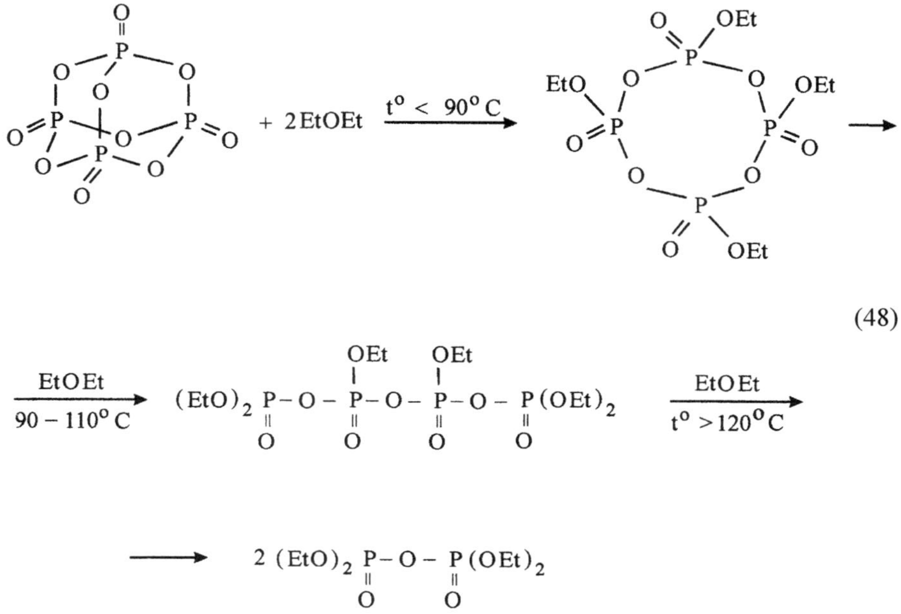

$$(48)$$

Hexaethyl tetraphosphate containing about 7–33% of tetraethyl meta-phosphate was claimed to be an active insecticidal agent [149].

Simultaneous with the above study [149] a similar investigation targeting the production of insecticides was carried out by Hull and Snodgrass [150–152]. It supported Adler's results [149], the interaction between phosphoric anhydride and diethyl ether upon heating for 8–12 h at 100–180°C producing mainly tetraethyl pyrophosphate. Though the possibility of tetraethyl pyrophosphate formation via tetraethyl metaphosphate and hexaethyl tetraphosphate [149, 152] (Scheme 48) was not ruled out, and another pathway to tetraethyl pyrophosphate was considered. This involved the reaction of phosphoric anhydride with triethyl phosphate which had been formed in the earlier stages of the process (Scheme 49).

$$P_4O_{10} + 6\,EtOEt \xrightarrow{100-180°C} 4\,(EtO)_3\,P{=}O \xrightarrow{0.5\,P_4O_{10}}$$

$$\longrightarrow 3\,(EtO)_2\,\underset{\underset{O}{\|}}{P}{-}O{-}\underset{\underset{O}{\|}}{P}(OEt)_2 \tag{49}$$

One of the main claims of the patent [152] was the pronounced insecticide activity of tetraethyl pyrophosphate, which was highly soluble in water and hydrolysed completely within 24 h. The good insecticidal properties of this compound were determined on 0.1–1.0% aqueous solutions, though it could be employed also in the form of dust or emulsion [152].

The patents issued for Hull and Snodgrass [150, 151] also covered the reaction of phosphoric anhydride with a variety of ethers including di-n-propyl, di-n-butyl, di-i-butyl and other symmetrical or asymmetrical alkyl, aryl or alkylaryl ethers (Scheme 50). In the earlier patent [150] examples were given only for the diethyl ether. The main objective for the study was the synthesis of trialkyl and triaryl phosphates in high yield and purity.

$$R\text{–}O\text{–}R^1 \xrightarrow{P_4O_{10}} (RO)_n (R^1O)_m P\text{=}O \tag{50}$$

$$R, R^1 = Et, Pr, n\text{-}Bu, i\text{-}Bu, Am, Ph, C_6H_4Me$$

$$n + m = 3$$

While a wide range of conditions for the process was covered, the principal methods of esterification used are as follows:

(A) Interaction of phosphoric anhydride with a large excess of anhydrous diethyl ether (reagent and solvent) in an autoclave at 130–140°C with or without a catalyst (iodine, cobalt acetate, potassium ferricyanide, sodium nitroprussid, ferric chloride, zinc chloride, concentrated sulphuric acid, phosphoric acid and a mixture of phosphoric acid with diphenyl ether). The maximum yield (75%) was achieved with catalysis by phosphoric acid at 146–150°C.

(B) Interaction of phosphoric anhydride with an excess of commercial diethyl ether that contained some water in the presence of phosphoric acid (100%) in an autoclave at 140–150°C for 15–48 h. The maximum yield of trialkyl phosphate in this case was 56%.

When an excess of diethyl ether was reacted with phosphoric anhydride in an inert solvent (benzene) at 150°C for 4 h, a 51% yield of triethyl phosphate was obtained. In all cases the products were isolated by vacuum distillation.

Unlike many other patents the reaction pathway was described as involving the steps shown in Schemes 51–53.

$$P_4O_{10} + 2R\text{–}O\text{–}R \longrightarrow (ROPO_2)_4 \xrightarrow{4R\text{–}O\text{–}R} 4(RO)_3P\text{=}O \tag{51}$$

$$P_4O_{10} + 4R\text{–}O\text{–}R \longrightarrow 2(RO)_4P_2O_3 \xrightarrow{2R\text{–}O\text{–}R} 4(RO)_3P\text{=}O \tag{52}$$

$$\left.\begin{array}{l} (RO)_4P_2O_3 \longrightarrow ROPO_2 + (RO)_3P\text{=}O \\ ROPO_2 \xrightarrow{R\text{–}O\text{–}R} (RO)_3P\text{=}O \end{array}\right\} \tag{53}$$

The actual formation of Langheld-ether was not discussed in the patent [152] though the authors mentioned that all the steps presented above (Schemes 51–53) are equilibrium reactions and for that reason the overall process for trialkyl phosphate cannot come to completion due to the presence of the intermediate compounds (including metaphosphates and pyrophosphates) in the reaction mixture. Nevertheless the residual mixture could be subjected to a further reaction with diethyl ether to produce more triethyl phosphate. The possibility of triethyl phosphate formation directly from the reaction intermediates was studied experimentally by Hull and Snodgrass [152].

In a thorough investigation of the reaction of phosphoric anhydride with diethyl ether, Burkhardt, Klein and Calvin [76] varied the experimental conditions widely, changing the time of the process and the ratio P_4O_{10}:EtOEt in order to optimize the experimental conditions for the phosphorylation of biologically important compounds [153, 154] (Chapter 5). The reaction products prepared under different conditions possessed variable n_D, density, viscosity, and signal intensities in ^{31}P NMR spectra. The data contradicted the results published earlier by Ratz and Thilo [79], Pollmann and Schramm [146] (see above). The ratio of phosphoric anhydride to diethyl ether was found to be the most significant factor determining the composition of the mixture of products. As presented in the paper [76] ^{31}P NMR spectra contain three broad signals in the areas $c. -13$, -27. and -43 ppm. The most marked differences in the signals of various reaction mixtures occurred in the high and low field parts of the spectra. Tetrametaphosphoric ester contains four magnetically equivalent phosphorus atoms (P_{middle}) but isotetrametaphosphoric ester contains one terminal phosphorus atom ($P_{terminal}$), one branched phosphorus atom ($P_{branched}$), and two middle phosphorus atoms (P_{middle}). Thus in the latter case the ^{31}P NMR spectrum should contain two signals of equal intensities ($P_{terminal}$ and $P_{branched}$) and one signal of greater intensity (P_{middle}). Nevertheless the data presented in the paper [144] are 25% ($P_{terminal}$) and 3% ($P_{branched}$) and in publication [76] one can find the figures 6–13% and 15–32% respectively. Such results were explained by a partial hydrolysis of the metaphosphate cycles. Burkhardt and co-authors [76] employed ^1H NMR spectroscopy to follow the rate of hydrolysis of Langheld-ester and found only traces of hydroxyl groups formed under the experimental conditions. Hence an insignificant amount of hydrolysis was occurring when the reaction was performed under conditions minimizing the contact with moisture. The ratio of integrated intensities of signals OH:C_2H_5 was determined by ^1H NMR for different products. Nevertheless the spectral analysis of all mixtures produced [76] indicated the presence of $P_{terminal}$ (42.4%, i.e. 5.2) and $P_{branched}$ (22.9%, i.e. 1.8). The molecular weight of Langheld-ester determined by different researchers is also inconsistent. For example it corresponded to two residues of metaphosphate ester according to [141] and to six residues according to [140]. A later determination of Langheld-ester molecular weight made by ebullioscopic method in chloroform [146] supported a tetrameric structure of metaphosphate. The authors of [76] obtained data corresponding to the hexameric structure for some of the products but the same species indicated tetrameric structures in solvents that minimized association. This suggests that in chloroform the components of Langheld-ester can associate.

Eventually on the basis of the above results the authors [76] concluded that the mixtures of products consist of both cyclic and acyclic compounds: i.e. cyclic tetrametaphosphate ester (45–50%), cyclic isotetrametaphosphate ester (25–36%), and open-chain tetrapolyphosphate ester (14–30%). Along with the processes of addition and step-by-step P–O–P bonds' cleavage, the possibility of the disproportionation of components in the reaction mixture such as those described in the publication [155] for triethyl phosphate and phosphoric anhydride (Section 4) is discussed. Many weeks' storage of the products in the water-tight containers did not show any signs of the formation of pyrophosphoric esters [76].

The ^{1}H and ^{31}P NMR data determined by Van Wazer and Norval [156] (Table 3.1) for the reaction of diethyl- and dimethyl-ether with phosphoric anhydride in chloroform are comparable with the corresponding NMR spectra presented in the publication [76]. The authors of [156] emphasized that chloroform extracted from the initially formed gummy layer contained trialkyl phosphate and tetraalkyl pyrophosphate. The equilibrium mixtures of the reaction contained mainly linear polyphosphoric esters and numerous attempts to identify the corresponding cyclic products showed few to be present.

Schramm and Berger [157] found that the nature of substituents (R) influenced the ^{31}P NMR shifts in the polyphosphate structures or in other words the NMR spectra are dependent upon the nature of an ether that reacts with phosphoric anhydride (Tables 3.1, 3.2).

The influence of moisture upon the interaction of phosphoric anhydride with diethyl ether in benzene has also been studied [158].

1.4 REACTIONS WITH OXIDES

In 1951 the reaction of phosphoric anhydride with various olefin oxides was considered by Woodstock [136] as involving the insertion of organic groups into the P–O–P bonds of P$_4$O$_{10}$ tricyclic structure in accordance with Scheme 54.

Table 3.1 ^{31}P NMR data (δ, ppm) for polyphosphate esters

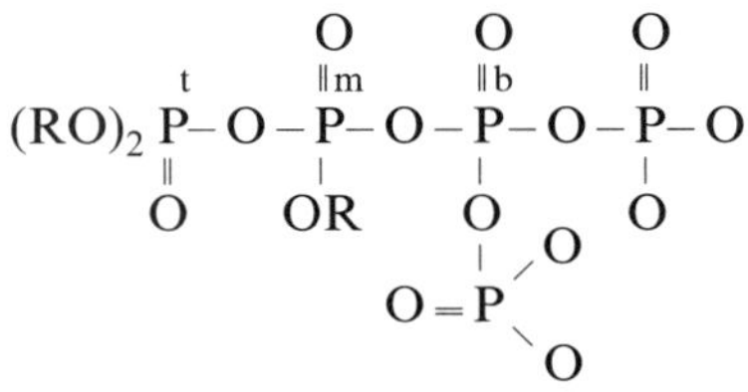

$$(RO)_2 \overset{t}{\underset{\overset{\|}{O}}{P}}-O-\overset{\overset{\overset{O}{\|}}{m}}{\underset{\underset{\displaystyle O=P\overset{/O}{\underset{\backslash O}{}}}{OR}}{P}}-O-\overset{\overset{\overset{O}{\|}}{b}}{\underset{O}{P}}-O-\overset{\overset{O}{\|}}{P}-O$$

R	P^t	P^m	P^b	Ref
Et	−16.0	−29.0	−42.0	[76]*
Et	−15.0	−29.0	−41.0	[156]*
Ph	−26.0	−35.0	−45.5	[157]

*The data of publications [76, 156] and [145, 144] are in close agreement

Table 3.2 ^{31}P NMR data for polyphosphates and phosphoric esters

Compound	δ (ppm)	Ref
1,5-μ-Oxotetrapolyphosphate (P_4O_{11})	−36.5	[8]
Tetrametaphosphate (P_4O_{12})	−23.6	[8]
Polyphosphates ('middle' P atoms)	−21.5	[8]
Polyphosphates (terminal P atoms)	−6-(−11)	[109]
Phosphoric acid (monoesters)	−3-(+4)	[109]

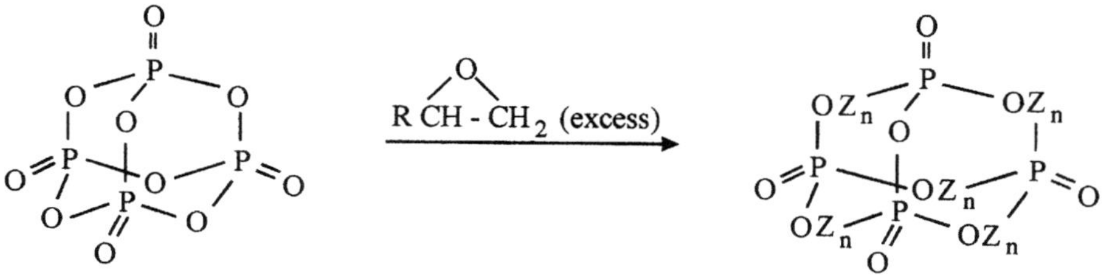

$$R = H, Me, Et, CH_2Cl, CH_2OH \; ; \qquad Z = \underset{R}{CH}-CH_2O \tag{54}$$

A large molar excess of olefin oxide (three to sevenfold) was recommended for this synthesis in the patent [136]. It was postulated that the caged structure shown in Scheme 54 could be stable until two molecules of an olefin oxide are bound to at least one phosphorus atom. It was suggested that breakdown of one or more bridging bonds resulted in the process of polymerization. Such speculations did not limit the author's view on the issue because he presumed some monomeric compounds of an undetermined structure were formed. The synthesis was performed in chloroform media with or without heating. In some cases the product polymerized at 100–130°C. According to elemental analysis the corresponding structures were evaluated as products of insertion of olefin oxides in the P–O–P bridge and polymerization [136] (Scheme 55), the actual structure depending on the molar ratio of olefin oxide–phosphoric anhydride.

$$(55)$$

In the case of epoxychlorohydrin no process of polymerization was observed even at 140°C [136]. Ethylene oxide exhibited a higher reactivity towards phosphoric anhydride than the substituted analogues. For example, isobutylene oxide reacted with phosphoric anhydride with a low yield and isopropylene oxide did not react at all. The products obtained in the above processes (both monomers and polymers) are of numerous valuable applications (Section 4, Chapter 6).

A detailed study of this reaction was performed by Wilcox, Harris and Olson [158a]. The recommended procedure for the reaction was very simple: ethylene oxide or substituted ethylene oxide was added in portions to a refluxing mixture of phosphoric anhydride and chloroform till the solid residue dissolved completely. On cooling the reaction mixture bisethylene pyrophosphate precipitated (27%). On removal of the solvent ethylene-bis(ethylene phosphate) was isolated (Scheme 56). The former product was moisture sensitive and hydrolysed violently into ethylene phosphoric acid [158a] (Scheme 56).

The authors [158a] considered this method to be a general synthetic approach to cyclic phosphates.

The preparation of various esters of phosphoric, pyrophosphoric and triphosphoric acids by the reaction of epoxides with phosphoric anhydride was also covered in the patent [159].

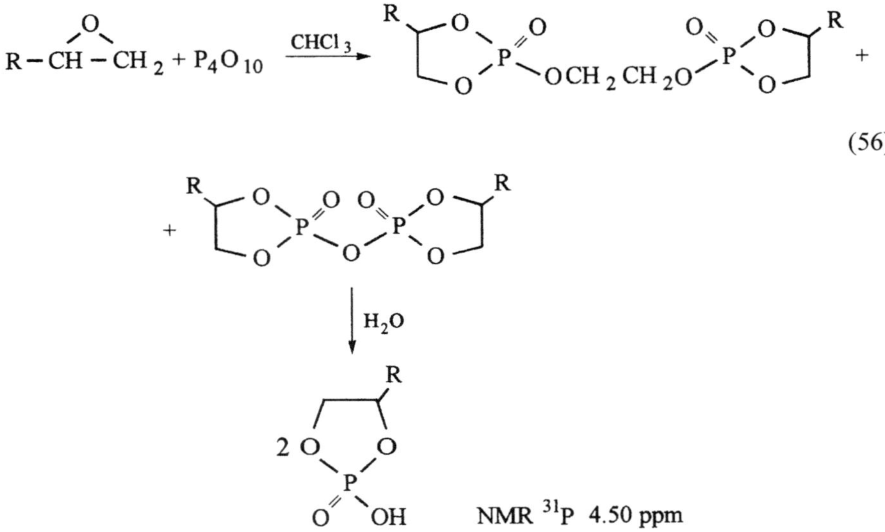

$$(56)$$

1.5 REACTIONS WITH ORTHOESTERS

In 1922 Staudinger and Rathsam [160] investigated the reaction of phosphoric anhydride with ethyl orthoformate, ethyl orthopropionate and ethyl orthoacetate with the aim of synthesising ketene acetal. The product was shown to be ethyl formate which was formed in accordance with Scheme 57 [160].

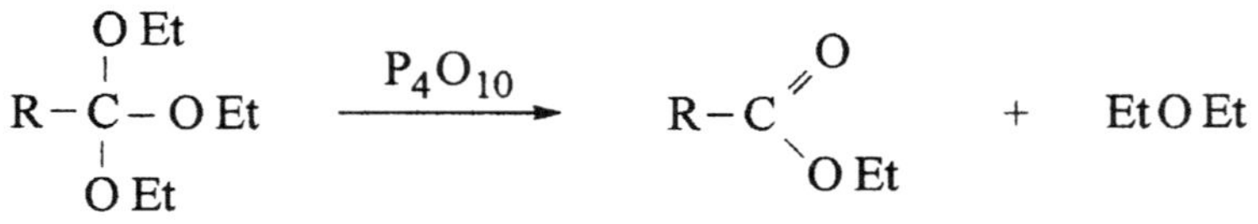

$$(57)$$

Diethyl ether was not isolated but this fact was explained by an interaction of ether with phosphoric anhydride under the reaction conditions. Brannock [161] found that under atmospheric pressure and relatively low temperature formation of Langheld-ester was very slow [138–140]. It was shown [161] that the reaction of P_4O_{10} with orthoesters proceeded at high temperature (125°C, 7 h) with the formation of either tetraethyl pyrophosphate and ethyl formate or triethyl phosphate and ethyl formate depending upon the ratio of the initial compounds (Schemes 58, 59).

$$P_4O_{10} + 4\,HC(OEt)_3 \longrightarrow 4\,HC(=O)OEt + 2\,(EtO)_2P(O)-O-P(O)(EtO)_2 \quad (58)$$

$$P_4O_{10} + 6\,HC(OEt)_3 \longrightarrow 6\,HC(=O)OEt + 4\,(EtO)_3P=O \quad (59)$$

As well as the reaction of phosphoric anhydride with orthoesters Brannock [161] studied the interaction of the same organic compounds with phosphoric thioanhydride. In the latter case alkylation was also accompanied by oxygen–sulphur exchange (Scheme 60).

$$HC(OC_2H_5)_3 + P_4S_{10} \longrightarrow HC(=S)SC_2H_5 + C_2H_5O\!-\!P(=S)(OC_2H_5)(SC_2H_5)$$

$$(60)$$

$$6\,HC(SC_2H_5)_3 + P_4S_{10} \longrightarrow 6\,HC(=S)SC_2H_5 + 4\,(C_2H_5S)_3P=S$$

1.6 REACTIONS WITH ESTERS

Reaction of phosphoric anhydride with malonic ester has been known for over 90 years [162]. It involves the processes of dealcoholysis (Section 3.8.2) leading to the isolation of the only product, carbon suboxide, with the yield less than 30% (Scheme 61).

$$EtOOC-CH_2-COOEt \xrightarrow[-EtOH]{P_4O_{10}} O=C=C=C=O \quad (61)$$

Despite numerous publications on the subject there were no descriptions of attempts to analyse the bulky solid residue. The authors of the current book extended the reaction to a study of phosphoric anhydride with mono-substituted acetic esters [163, 164]. The process leads to the formation of products of dealcoholysis i.e. corresponding monosubstituted ketenes (Section 8.2) and an investigation of the residual solids led to the isolation of acylphosphates and phosphorylated phloroglucinols. The reaction of P_4O_{10} with monosubstituted acetic esters and malonic ester is presented in detail in Section 8.2.

1.7 REACTION WITH ACETIC ANHYDRIDE

In 1921 Steinkopf and Schubart [142] acetylated thiophen with a mixture of phosphoric anhydride and acetic anhydride. The reaction was depicted as proceeding via acetylmetaphosphoric ester as shown (Scheme 62), the phosphorus group providing an improved leaving group.

$$P_2O_5 \ + \ Ac_2O \ \longrightarrow \ 2 \ \overset{O}{\underset{O}{\diagup}}P-OAc$$

$$\tag{62}$$

$$\overset{O}{\underset{O}{\diagup}}P-OAc \ + \ \underset{S}{\bigcirc} \ \longrightarrow \ \underset{\overset{+}{S}-Ac}{\bigcirc} \ + \ PO_3^-$$

$$PO_3^- \ + \ \overset{O}{\underset{O}{\diagup}}P-OAc \ \longrightarrow \ P_2O_5 \ + \ AcO^-$$

A reinvestigation of the reaction pathway using modern spectroscopic methods should be worthwhile.

2 REACTION WITH THIOLS

Mercaptotetrazoles were rather inert towards phosphoric anhydride but the reaction of their sodium salts with P_4O_{10} and further treatment by dilute hydrochloric acid gave the corresponding mercaptotetrazolyl pyrophosphates in low yield [165] (Scheme 63).

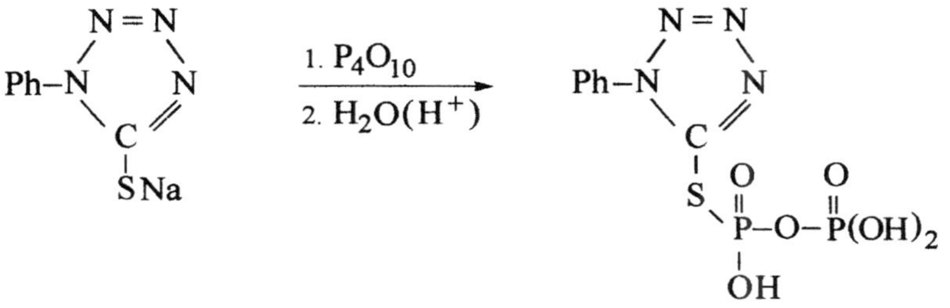

$$
\text{(63)}
$$

For the reactions of phosphoric anhydride with acyclic and cyclic thioamides see Sections 3.3 and 3.4.

3 REACTIONS OF PHOSPHORIC ANHYDRIDE WITH NITROGEN-CONTAINING COMPOUNDS

3.1 REACTIONS WITH AMMONIA

Phosphoric anhydride reacted vigorously with liquid ammonia in a thick tube at room temperature [3, 166, 167]. A stable product was formed after 4–6 weeks. It was assigned structure (18) by Becke-Goering and Sambeth [166].

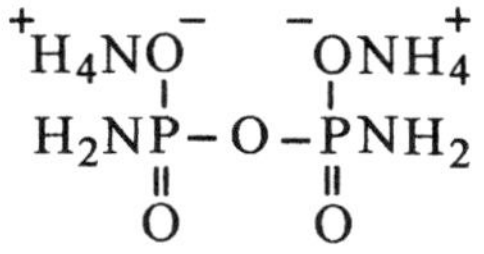

$$
\overset{+}{\text{H}_4}\text{NO}^- \qquad {}^-\text{ON}\overset{+}{\text{H}_4}
$$

(18)

At 40–50°C the stable product was formed in a rotating autoclave within 24 h. With longer heating at 100–110°C, ammonia was evolved and a brown product of polycondensation was formed (P : N$_{bonded}$: N$_{ammonia}$ = 1 : 1.5 : 1.06) [166] (Scheme 64). Under some conditions a mixture of phosphoric and pyrophosphoric acid amides and their ammonia salts may be formed (Scheme 65).

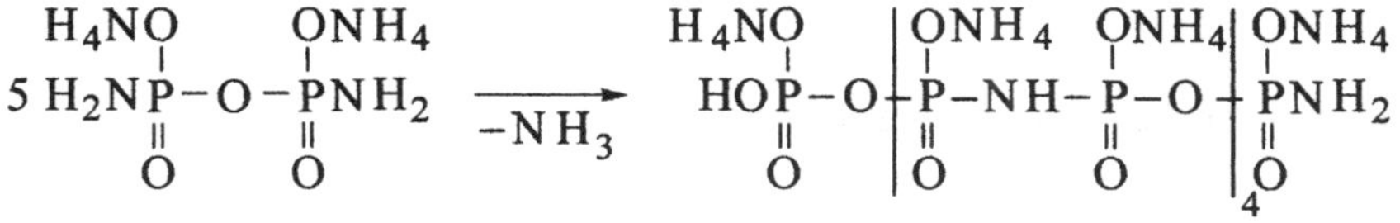

$$
\text{(64)}
$$

$$NH_3 \; + \; P_4O_{10} \longrightarrow$$

$$\tag{65}$$

The products of the reaction of phosphoric anhydride with ammonia in the presence of water were open-chain phosphates containing the amide and imide groups [3, 168–173]. The general procedure involved vapour phase mixing of phosphoric anhydride, ammonia and water [168]. The reaction rate was very fast and the product condensed upon cooling the reaction product. The nature of the products was influenced significantly by the ratio of the initial reagents, temperature of P_4O_{10} vapour, time of contact and period of condensation [168]. Usually the reaction gives a complicated mixture of products. Van Wazer [3] suggested a correlation between the structure of products formed and the ratio of the starting materials ($P_4O_{10}:NH_3:H_2O$). The typical polyphosphate structure was assigned [3] as having two triphosphate fragments combined by a bridging N–H group.

Ammonium polyphosphate is manufactured by heating P_4O_{10} with ammonium phosphate in an atmosphere of ammonia [174, 175].

3.2 REACTIONS WITH AMINES

Interaction of phosphoric anhydride with primary and secondary amines follow the general scheme (Scheme 66) for the reaction between P_4O_{10} and nucleophilic reagents as presented above.

Buck and Lankelma [176] described the reaction of phosphoric anhydride with aniline in toluene at 50°C, the following treatment with sodium hydroxide solution (5%) gave bis (phenylamido) phosphoric acid (Scheme 67). The authors [176] suggested that HPO_3 did not phosphorylate aniline but formed only its salt.

The principal publication concerning reactions of phosphoric anhydride with amines was presented by Cherbuliez and co-workers [96]. It was found out that (a) the reactivity of phosphoric anhydride towards amines decreases with diminishing of the P–O–P bond number similar to the reactions with alcohols (Section 1.2); (b) the amines of lower basicity (for example aniline) give products of higher aminolysis of phosphorus than the more basic amines such as Et_2NH (Scheme 68).

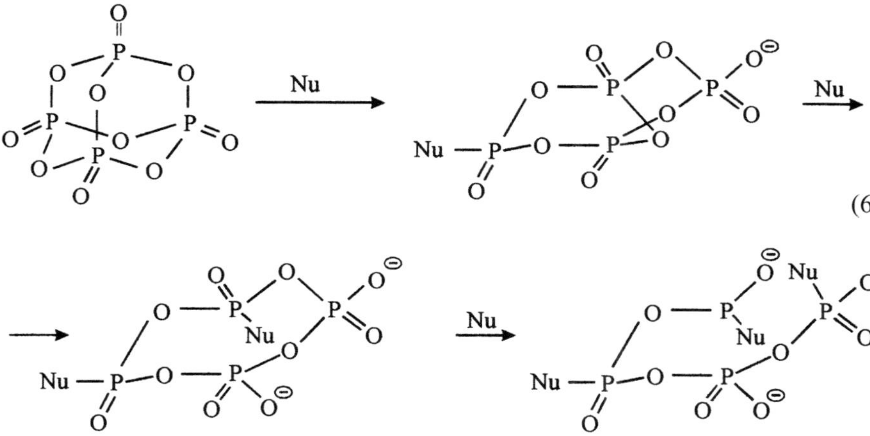

$$(66)$$

$$P_4O_{10} + 4\,PhNH_2 \longrightarrow 2\,(PhNH)_2\,P(O)\text{--}OH + 2\,HPO_3 \tag{67}$$

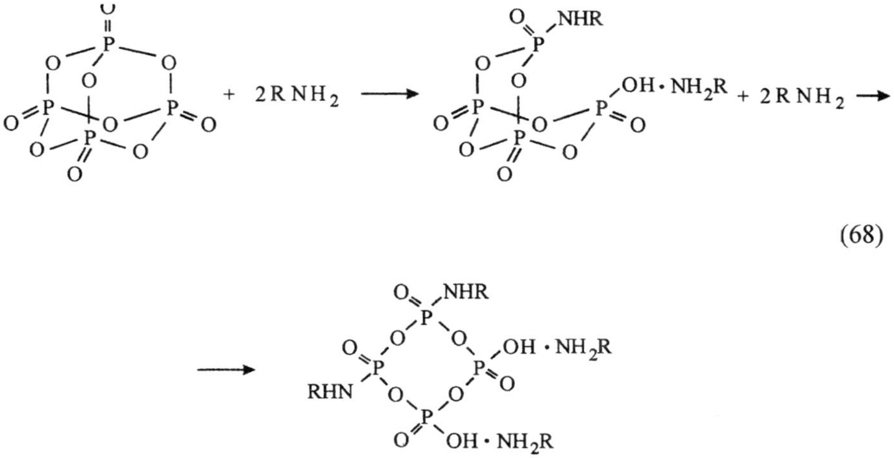

$$(68)$$

The nature of products formed from phosphoric anhydride and secondary amines depends upon the initial reagents' ratio [96] (Schemes 69, 70). The spectroscopic characteristics of phosphorylated amines were not presented [96].

Wolf and Meisel [177] investigated the interaction between phosphoric anhydride and pyridine which leads to the formation of a complex compound of the type $Py_2 \cdot P_2O_5$ [177]. The reaction of this product with phosphoryl trichloride and pyridine was investigated (Schemes 71–74).

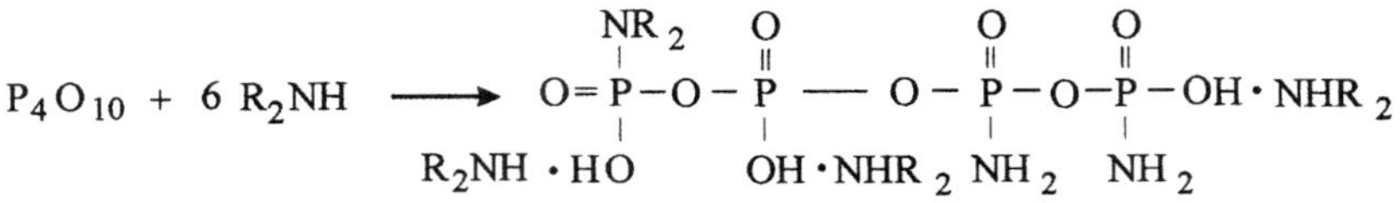

$$P_4O_{10} + 6\,R_2NH \longrightarrow \underset{\underset{R_2NH \cdot HO}{|}}{\overset{\overset{NR_2}{|}}{O=P}}-O-\underset{\underset{OH \cdot NHR_2}{|}}{\overset{\overset{O}{\|}}{P}} - O - \underset{\underset{NH_2}{|}}{\overset{\overset{O}{\|}}{P}}-O-\underset{\underset{NH_2}{|}}{\overset{\overset{O}{\|}}{P}}-OH \cdot NHR_2 \qquad (69)$$

$$P_4O_{10} + 12\,R_2NH \longrightarrow 2\,R_2N-\underset{\underset{O}{\|}}{P}(O^-\,R_2\overset{+}{N}H_2) + 2(R_2N)_2\underset{\underset{O}{\|}}{P}\,O^-\,R_2\overset{+}{N}H_2 \qquad (70)$$

$$P_4O_{10} + 4\;\text{[pyridine]} \longrightarrow 2\,(\text{[pyridine]}N)_2 \cdot P_2O_5 \qquad (71)$$

$$(\text{[pyridine]}N)_2 \cdot P_2O_5 + POCl_3 \xrightarrow{\;\text{[pyridine]}\;} 3\,(\text{[pyridine]}N) \cdot PO_2Cl \qquad (72)$$

$$\text{[pyridine]}N \cdot PO_2Cl + H_2O \xrightarrow{\;\text{[pyridine]}\;} \text{[pyridine]}N \cdot PO_2OH + \text{[pyridine]}N \cdot HCl \qquad (73)$$

$$\text{[pyridine]}N \cdot PO_2OH + \text{[pyridine]}N \cdot PO_2Cl \xrightarrow{\;\text{[pyridine]}\;} (\text{[pyridine]}N)_2 \cdot P_2O_5 + \qquad (74)$$

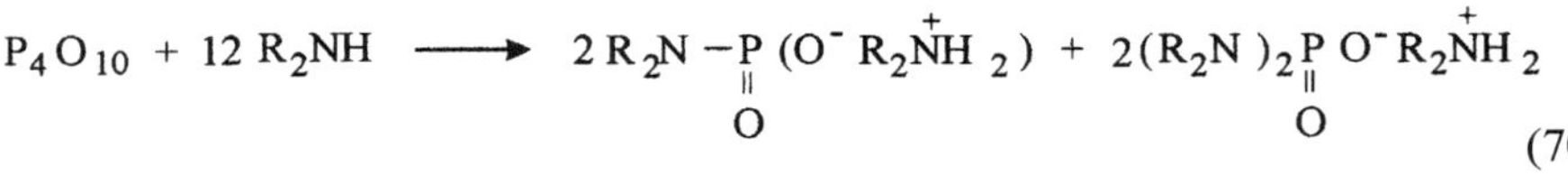

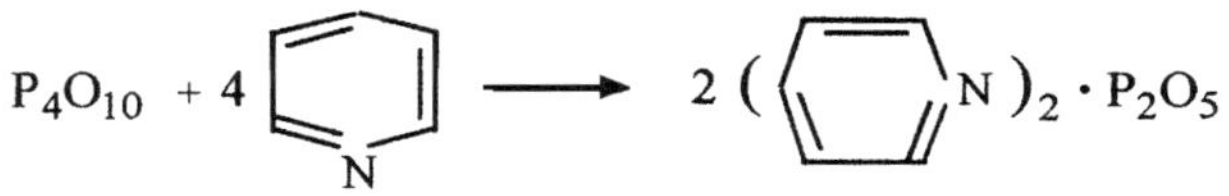

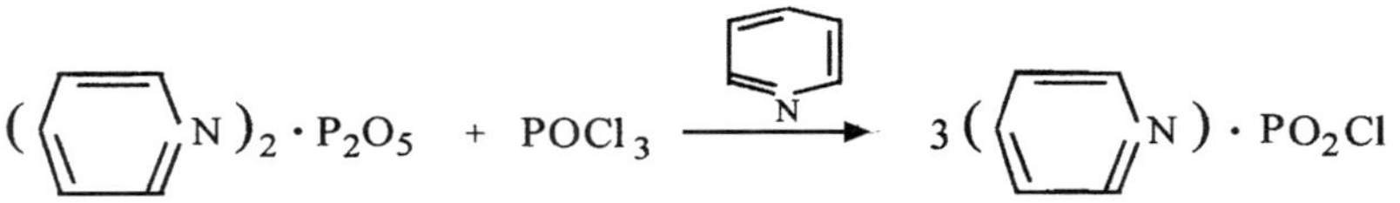

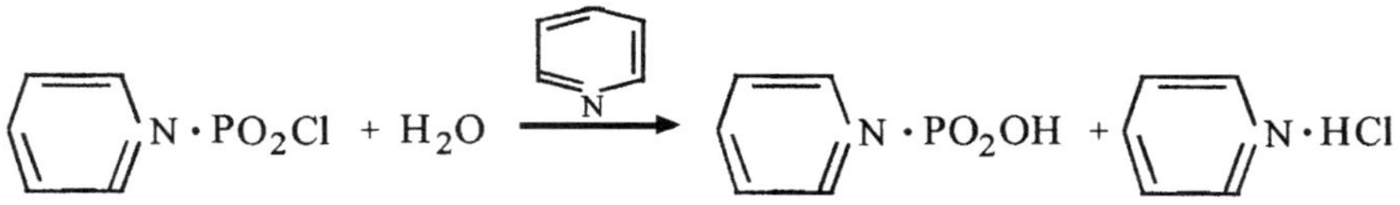

$$+ \text{[pyridine]}N \cdot HCl$$

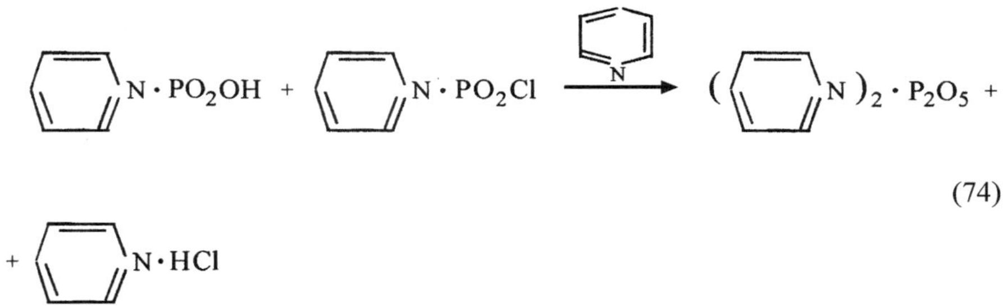

The reactions of amines with phosphoric thioanhydride proceed further than with phosphoric anhydride. For example P$_4$S$_{10}$ reacted with aqueous ammonia to give ammonia dithiophosphates (Scheme 75).

$$P_4S_{10} + 12\,NH_3 + 8\,H_2O \longrightarrow 4(NH_4)_3PO_2S_2 + 2\,H_2S \tag{75}$$

In the case of gaseous and liquid ammonia the following thiophosphorylamide and iminophosphoryl compounds were formed [178] (Scheme 76).

$$P_4S_{10} + NH_3\text{ (gas)} \longrightarrow (NH_4S)_2P\underset{\underset{S}{\|}}{N}H_2 + (NH_4S)_3P{=}S \tag{76}$$

$$P_4S_{10} + 12\,NH_3\text{ (liq.)} \longrightarrow 2\ \overset{\overset{SNH_4}{|}}{\underset{\underset{SNH_4}{|}}{HN{=}P}}{-}S{-}\overset{\overset{SNH_4}{|}}{\underset{\underset{SNH_4}{|}}{P{=}NH}}$$

The reaction of P$_4$S$_{10}$ with aliphatic primary amines (molar ratio 1:8) led to the formation of diamidodithio phosphates whereas the excess of primary amines directed the process towards triamido thiophosphates with the yield about 50% [179, 180] (Scheme 77).

$$P_4S_{10} + 8\,AlkNH_2 \longrightarrow 4\,(AlkNH)_2PSSH + 2\,H_2S$$
$$P_4S_{10} + 12\,AlkNH_2 \longrightarrow 4\,(AlkNH)_3PS + 6\,H_2S \tag{77}$$

The dianilide of dithiophosphoric acid was synthesized by the same method [181] (Scheme 78).

$$P_4S_{10} + 8\,H_2NC_6H_5 \longrightarrow 4(C_6H_5NH)_2PSSH + 2\,H_2S \tag{78}$$

Wise and Lankelma [182] also studied the reaction of P$_4$S$_{10}$ with secondary amines which usually gave a mixture of substituted thiophosphoric acids (Scheme 79).

$$P_4S_{10} + 6(C_4H_9)_2NH \longrightarrow 2(C_4H_9)_2NP\underset{\underset{S}{\|}}{(SH)_2} + 2[(C_4H_9)_2N]_2PSSH \tag{79}$$

3.3 REACTIONS OF ACYCLIC AMIDES AND THIOAMIDES

It is impossible to overestimate the great significance of amides in the evolution of life on Earth. Amides were among the earliest organic compounds formed and presumably were phosphorylated by primary phosphorylating agents.

Twenty years ago, a hypothesis was presented concerning the participation of phosphoric anhydride in the processes of phosphorylation on the primitive Earth [183] (Chapter 4).

Phosphorylated di and tripeptides possess a high activity against Gram-negative bacteria [184]. Another valuable feature of phosphorylated amides is their insecticide activity. For example O,S-dimethyl-N-acetylamidothio phosphate ('Acephat') is an efficient insecticide with a wide range of activity and low toxicity [185, 186].

The C(O)–N bond in amides possesses a partially double character which influences on stereochemical [187–190], chemical and spectroscopic [191] properties of amides. In this respect the authors of the current book conducted a study of the reactions of phosphoric anhydride with mono and disubstituted amides. The attractive points for such a study were the presence of two competing nucleophilic centres in amides and their potential to tautomerise which could have a pronounced role in their reactions with P_4O_{10} (Scheme 80).

$$
\begin{array}{ccc}
R-N\!\!\begin{array}{c}H\\[-2pt]C-R'\\[-2pt]\parallel\\[-2pt]O\end{array} & \rightleftharpoons & R-N=C\!\!\begin{array}{c}R'\\[-2pt]OH\end{array}
\end{array} \qquad (80)
$$

a   b

3.3.1 Reactions with benzoylamides and thiobenzoylamides

The reactions of phosphoric anhydride with benzoylamide and thiobenzoylamide were carried out under reflux in chloroform or carbon tetrachloride for 18–24 h [192]. The following mild treatment of a viscous residue with water resulted in isolation of mixtures of N and O (S) phosphorylated amides, N-benzoylamido phosphate and benzimidoyl phosphate from the former amide, N-thiobenzoylamido phosphate and thiobezimidoyl phosphate from the latter one (Scheme 81). According to ^{31}P NMR the ratio of isomers in different solvents (acetone-d_6, DMSO-d_6, ethanol-d_6) varied within narrow limits.

Earlier the possibility of the formation of O-phosphorylated amides in biological systems had been proposed by Grout [193]. This related to the products of the phosphorylation of glycinamidoribotide by ATP although it had not been proposed by the authors of the original publications [194, 195].

$$(81)$$

X = O, S
ⓟ = Polyphosphate

^{31}P NMR monitoring of the reaction of phosphoric anhydride with benzoylamides [192] allowed the site of the primary attack by P$_4$O$_{10}$ to be deduced which was the more nucleophilic nitrogen atom of the amides or their imidol tautomers [191]. This was followed by the process of N$\rightarrow$O (S) migration of the phosphoryl group (phosphorotropy) under the action of moisture or protic solvents. The 1–3 shift was more pronounced in amido phosphates than in thioamido phosphates [192] due to the higher stability of the P–O bond over the P–S bond.

The effect of 1–3 migration of the phosphoryl group and the term 'phosphorotropy' were first presented by Zimin, Afanasiev and Arkady Pudovik in 1979 [196] (See Section 3.3.3). In their study benzoylamides had been phosphorylated predominantly by alkylchloro phosphates, phosphorus pentachloride and some other reagents that produced fully esterified dialkyl (diaryl) amido phosphates. Synthesis and properties including biological activity of N-phosphorylated arylamides and thioarylamides complexes with different metals have been studied in depth by Kazan scientists [197–202]. Biology-related phosphorus (III) containing amides are described in the publications presented by Nifantyev and Gratchev [185, 203].

3.3.2 Reactions with acetanilides

Acetanilides and their derivatives are the objects of particular attention of chemists as well as biologists, pharmacologists and environmental scientists due to a rich variety of their applications. Acetanilides derivatives are common components of drugs [204, 205]. The derivatives of acetanilide that have been widely used in agriculture, such as alachlor, acetchlor, butachlor, metolachlor, were found to be, not only very active herbicides, but also extremely toxic compounds with mutagenic and some other undesirable biological properties that initiated a detailed study of their activity [206–214].

In their reactions towards phosphoric anhydride acetanilides were much more reactive than benzoylamides (Section 3.3.1). Refluxing of these compounds in chloroform for 1–3 h resulted in the formation of phosphorylated products which reacted readily with further amide to give crystalline amidines [215b]. The primary attack by P_4O_{10} was directed towards the nucleophilic nitrogen atom of amides [215–218] (Scheme 82).

$$(82)$$

$$R, R^1, R^2 = H, Alk \; ; \quad \textcircled{P} = Polyphosphate$$

Under the action of moisture or protic solvents the primary products isomerized partially into the O-phosphorylated form. The consequence of processes in phosphorylated acetanilides was influenced by the ratio of the initial reagents. An equilibrium between N and O-phosphorylated isomers took place [218] in the case of the molar ratio P_4O_{10}:PhC(O)NHMe 1:1–2. In all mixtures O-phosphorylated isomers were prevailing. The methyl substituents in the aromatic ring enriched its electronic density and slowed down the process of N$\rightarrow$O migration. Computer modelling (AM1 and PM3) of both types of products showed the data E_{total} and $\Delta H°_f$ favouring N-arylacetimidoyl phosphoric acids (O-phosphorylated derivatives).

The process of phosphorylation and the following N$\rightarrow$O migration of the hydroxyphosphoryl group (phosphorotropy) was monitored by ^{31}P NMR spectroscopy (Figure 3.2) which exposed the intermediates and the products for all acetanilides (Table 3.3). The time-dependent spectra of asymmetrically substituted N-acetyl-(2-methylphenyl) amide and symmetrically substituted

N-acetyl-(2,6-dimethylphenyl) amide are presented in the corresponding figures of the publication [218]. The initially registered group of signals in the phosphorus-31 NMR spectra centred at -40 ppm, which is characteristic for the branched phosphorus atoms in the bicyclic structure formed as the result of the first step of the nucleophile addition to tricyclic phosphoric anhydride were soon replaced by two complex groups of signals in the regions from -28 to -25 ppm and from -13.5 to -11.5 ppm (Figure 3.2 'a' and 'b'). The former signals were attributed to polymetaphosphoric acid whereas the low field signals were assigned to pyrophosphates and N-phosphorylated acetanilide in the reaction mixture. In the course of time the P–O–P bonds were partially split by the organic reagent which was reflected by simplifying the groups of signals between -28 and -25 ppm and 13.5 and -11.5 ppm as is shown by Figure 3 'c'. The strong high-field singlet at -24 ppm was assigned to the presence of metastable trisubstituted trimetaphosphate.

The addition of traces of hydrogen chloride resulted in a growth of the new signal registered by ^{31}P NMR in the region from -2 to 3 ppm (Figure 3 'd' and 'e'), which is characteristic for the monosubstituted phosphoric acid. These changes were attributed to N→O migration of the hydroxyphosphoryl group. Simultaneously the higher field signals transformed into one less intensive signal. Within 90–270 min the ratio between the signals attributed to the N- and O-phosphorylated acetanilides stabilized, suggesting that an equilibrium between these two isomers had been established (Figure 3.2 'g', Table 3.3). Further hydrolysis of phosphorylated acetanilides did not take place. According to Gasteiger and co-authors [219] the rate of amide hydration at pH 7 is $c.$ 1000 times higher than in acidic media. Hence the acidic dihydroxyphosphoryl group did not promote a process of product degradation.

In the crystalline forms of the product under anhydrous conditions, phosphates were usually prevailing over amidophosphates. This was determined by solid-state ^{31}P NMR but under the action of atmospheric moisture or in solutions of polar solvents the equilibrium appeared to be shifted partially towards the N-phosphorylated acetanilides [218]. Acidity of the medium also influenced the equilibrium between the N- and O-phosphorylated isomers. Under the influence of an acid the equilibrium shifted to the phosphate and on addition of a base to the product partial O→N phosphorotropy [218]. The signals attributed to amidophosphates were enhanced by the presence of methyl substituents in the aromatic ring which was particularly marked in N-acetyl- N-(2,4,6-trimethylphenyl) amidophosphate.

Molecular modelling of the pairs of structures by the programs AM1 and PM3 (MOPAC 6) exhibited a small energy difference between the two forms with the values E_{total} and ΔH_f^0 favouring the O-phosphorylated acetanilides over the N-phosphorylated compounds [218].

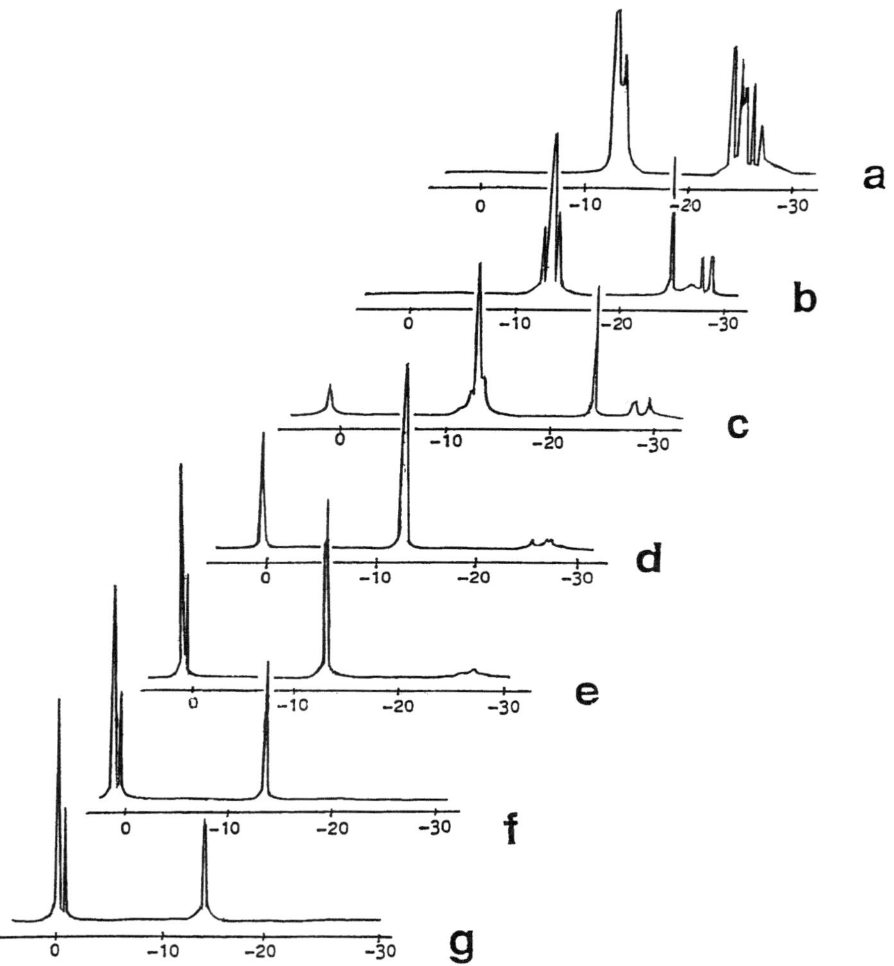

Figure 3.2 Time-dependent phosphorus-31 NMR spectra of the reaction mixture P_4O_{10}—N-(2,4,6-trimethylphenyl) acetanilide [Authors]

Formation of the O-phosphorylated structure was confirmed also by means of ^{13}C NMR [218] (Table 3.3). The spectra of the products contained either two or one broad signal in the region 143–151 ppm which correlated with the fragment $N=\underline{C}(Me)OP(O)(OH)_2$. In the case of amidophosphates substituted in the aromatic ring the imide carbon atom of this fragment was registered at the higher field.

Table 3.3 Spectroscopic characteristics of acetanilides and their phosphorylated derivatives

| | | | ^{31}P NMR, δ (ppm) | | ^{13}C NMR, δ (ppm) | | | IR, v (cm^{-1}) | | |
Type of compound	R^{ortho}	R^{para}	N–PO$_3$H$_2$	O–PO$_3$H$_2$	N–C=O	N=C–O	N–H	P=O	P–O–C	OH
I	H, H	H			166.67		3294			
II	H, H	H	−13.06		170.86			1224, 1218		3320
III	H, H	H		−1.44 br		141.08		1224, 1218	1028	3320
I	H, Me	H			166.46		3222			
II	H, Me	H	−13.33		169.68			1261, 1224		3400–3000
III	H, Me	H		−1.89		145.25		1261, 1224	1020 br	3280
I	Me, Me	H			166.11		3235			
II	Me, Me	H	−12.76		170.89			1257, 1217, 1199		3200

continued

Table 3.3 *continued*

Type of compound	R^{ortho}	R^{para}	^{31}P NMR, δ (ppm)		^{13}C NMR, δ (ppm)			IR, v (cm^{-1})		
			N–PO$_3$H$_2$	O–PO$_3$H$_2$	N–C=O	N=C–O	N–H	P=O	P–O–C	OH
III	Me, Me	H		1.27		137.95 138.09		1257, 1217, 1199	1014, 986	3200
I	Me, Me	Me			167.73		3200			
II	Me, Me	Me	−12.88		172.10			1270, 1222, 1210		3150
III	Me, Me	Me		2.71, 2.88		137.95 141.83		1270, 1222, 1210	1016, 973	3150
I	Pr^i, Pr^i	H			169.92		3272			
II	Pr^i, Pr^i	H	−13.31		167.52			1233, 1185		3390
III	Pr^i, Pr^i	H		−2.36, −2.89		147.17 147.36		1233, 1185	1060, 997	3390

The NMR spectra are registered in (CD$_3$)$_2$SO
The IR spectra are registered in KBr

It is noteworthy in ^{13}C and ^{1}H spectra that the position of the dihydroxyphosphoryl group did not influence significantly the chemical shifts of the signal of the methyl group in the acyl fragments [218]. In all derivatives, the CH$_3$ groups were magnetically equivalent as had been determined earlier [220].

Amides react in a nucleophilic manner with other electrophilic reagents and substrates such as alkyl sodides and cabonyl compounds. The latter reactions (Scheme 83) are well studied and known to be reversible in neutral and basic media [221].

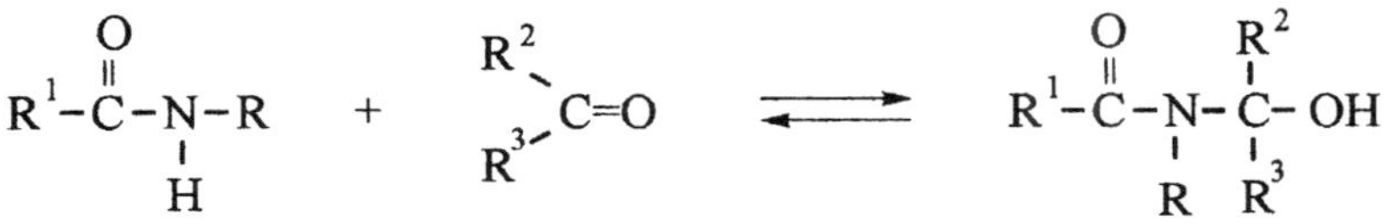

$$R^1\text{-}\overset{\overset{\displaystyle O}{\|}}{C}\text{-}\underset{\underset{\displaystyle H}{|}}{N}\text{-}R \quad + \quad \overset{R^2}{\underset{R^3}{}}C\text{=}O \quad \rightleftharpoons \quad R^1\text{-}\overset{\overset{\displaystyle O}{\|}}{C}\text{-}\underset{\underset{\displaystyle R}{|}}{N}\text{-}\overset{\overset{\displaystyle R^2}{|}}{\underset{\underset{\displaystyle R^3}{|}}{C}}\text{-}OH \qquad (83)$$

Phosphorylated acetanilides nonsubstituted in the acyl fragment did not exhibit any tendency to undergo intramolecular dehydration [215] apart from the intermolecular condensation reaction to form amidines (see Section 8.1).

The ^{13}C NMR spectra of the phosphorylated acetanilides were complicated by the formation of amidines, and the presence of isomers of chain (N and O-phosphorylated products) and stereomers E **(19)** and Z **(20)** of the O-phosphorylated acetanilides [218] (Table 3.3, Figure 3.2).

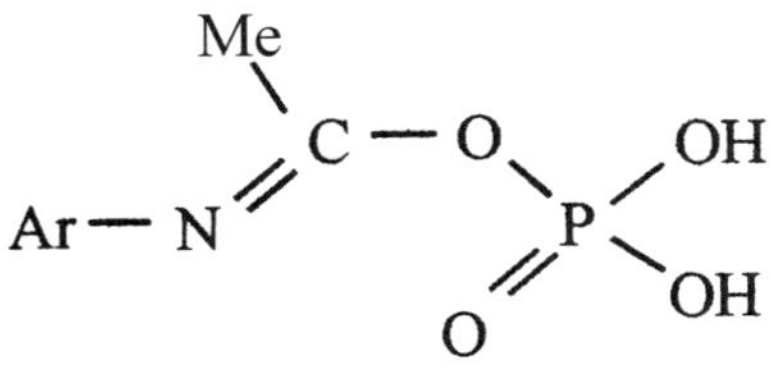

E   **(19)** 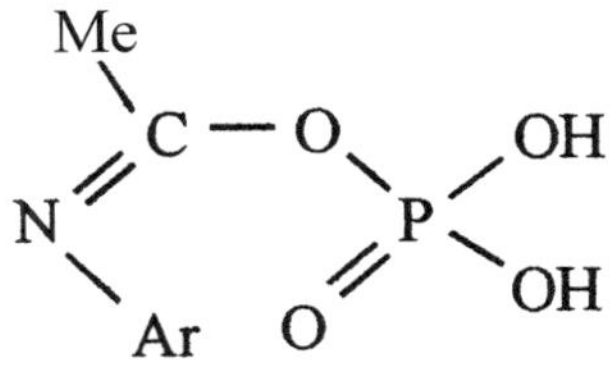   Z   **(20)**

A high ratio of acetanilides in the reaction with phosphoric anhydride (over 2 : 1) favoured the spontaneous interaction between phosphorylated acetanilides and unreacted amide. The dihydroxyphosphoryl (phosphate) group acted as a leaving group [222, 223] and the corresponding products of condensation 2-methyl-1,3-diphenyl-3-acetamidines were formed [218, 224, 225] (Scheme 84).

The mixture of N-(2,6-diisopropylphenyl)-N-acetamidine with N-acetyl-N-(2,6-diisopropylphenyl) amidophosphate was the most reactive species in the process of condensation due to the 'pushing' effect of the bulky diisopropyl substituents directed towards the readily leaving dihydroxyphosphoryl group.

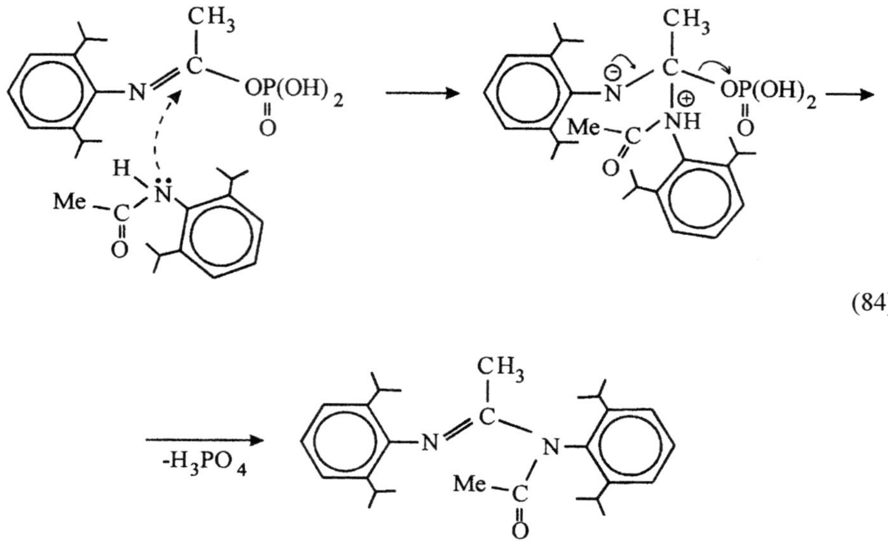

$$(84)$$

This process can be considered as a variation of amide condensation [226]; in this particular case promoted by the phosphate fragment.

3.3.3 Reactions with benzanilides

Benzanilides were more reactive towards phosphoric anhydride than the primary amides of benzoic acid (Section 3.3.1) and acetanilides (Section 3.3.2). The most reactive species were the dimethyl- and trimethyl-substituted benzanilide. The reaction of these compounds with phosphoric anhydride was spontaneous and exothermic, whereas the nonsubstituted and mono-methyl-substituted benzanilides were refluxed for about 3 h for the process to complete [227–230] (Scheme 85). The yields of the products were rather high (65–95%). The optimum molar ratio P_4O_{10}:benzanilide was determined to be 1:2 [229, 230].

Monitoring the intereaction between phosphoric anhydride and benzanilides by means of phosphorus-31 NMR showed a somewhat similar pattern of the time-dependent spectra as is shown in the Figure 3.3. The time-dependent ^{31}P NMR spectra reflecting the interaction of nonsubstituted benzanilide with phosphoric anhydride can be found in the publication [230]. Shortly after mixing the reagents (P_4O_{10} is insoluble in chloroform and carbon tetrachloride) the spectra showed a group of signals centred at about -40 ppm reflecting the beginning of the chemical process. This group of signals was assigned to the bridging phosphorus atoms in the bicyclic structure

[230] formed after the reaction with one molecule of benzanilide. Within 20–30 min there were no signals in the region between −36 and −44 ppm, instead two groups of signals in the regions from −25 to −27 and the other from −12 to −13 ppm appeared (Figure 3.3 'a' and 'b').

$$(85)$$

R, R^1, R^2 = H, Alk ;   Ⓟ = Polyphosphate

The higher field signals were assigned to the phosphorus atoms in metaphosphoric acid and the middles of tetraphosphoric acid [230]. The lower field group of signals reflected the end phosphorus atoms in tetraphosphoric and pyrophosphoric acids and the amidophosphate fragment. Magnetic nonequivalent (spin–spin coupling) phosphorus atoms typical for unsymmetrically substituted pyrophosphoric acids were not registered (Figure 3.3). Under the influence of atmospheric moisture or moist solvents the P–O–P bonds underwent slow hydrolysis reflected in the phosphorus-31 NMR spectra by the gradual decreasing set of signals centred at −26 ppm meanwhile the signals in the region from −12 to −13 ppm gained intensity while changing the pattern (simplifying) (Figure 3.3 'b' and 'c'). The solid moderately stable products registered in the spectra by a dominating signal at about −12 ppm (Figure 3.3 'd', Table 3.4) were isolated and determined to be the corresponding N-phosphorylated amides. IR spectra of these compounds did not contain the

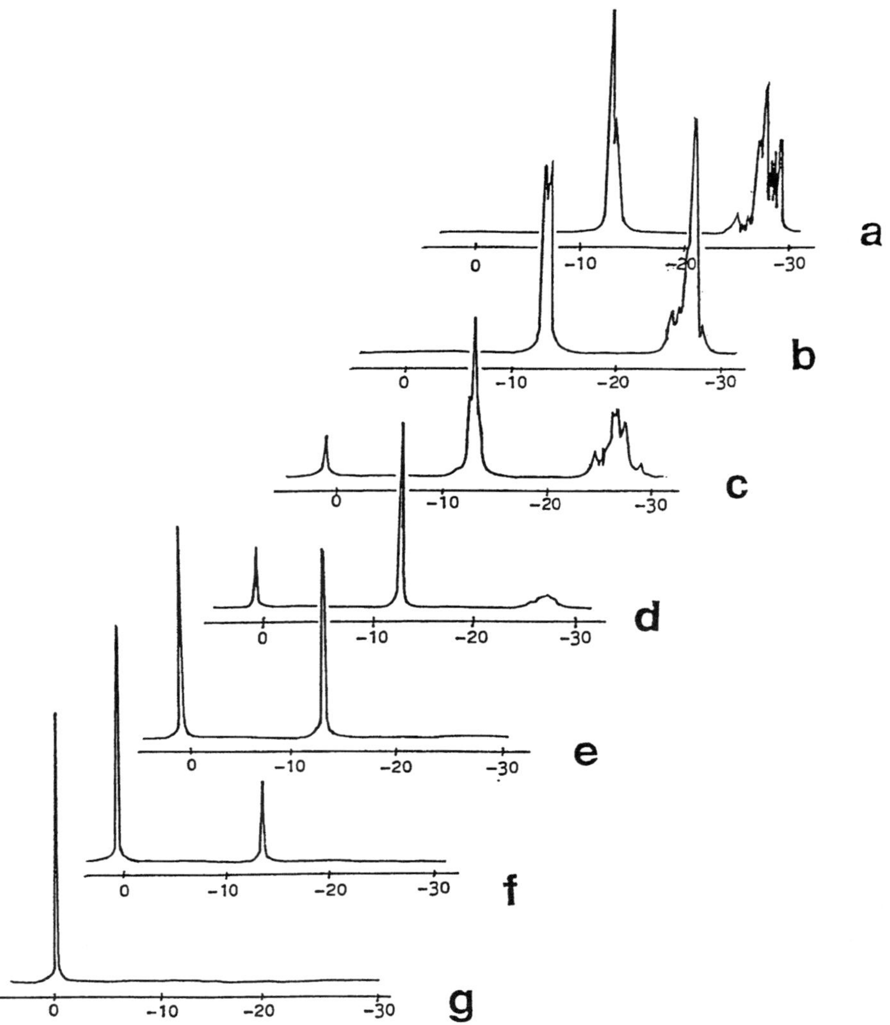

Figure 3.3 Time-dependent phosphorus-31 NMR spectra of the reaction mixture P_4O_{10} — N-(2,4,6-trimethylphenyl) benzanilide [Authors]

bands characteristic to the N–H bond in the amide group (about $3250\,\text{cm}^{-1}$) (Table 3.4) [230].

The signals of the carbonyl groups in ^{13}C NMR registered in the chemical shift region 163–173 ppm also confirmed the regioselectivity of the process (N-phosphorylation) (Table 3.4). Measuring of the $^{1}J_{\text{N-P}}$ value by means of ^{15}N

Table 3.4 Spectroscopic characteristics of benzanilides and their phosphorylated derivatives

Type of compound	R^{ortho}	R^{para}	^{31}P NMR, δ (ppm)		^{13}C NMR, δ (ppm)		IR, ν (cm^{-1})			
			N–PO$_3$H$_2$	O–PO$_3$H$_2$	N–C=O[1]	N=C–O[1]	N–H	P=O	P–O–C	OH
I	H, H	H			166.72 [166.50[2]]		3290			
II	H, H	H	−13.6		171.25 [173.43[2]]			1255		3300– 300
III	H, H	H		2.81, 3.70		142.3 [141.43[2]]		1275, 1255, 1200	1010, 1000	3100– 2900
I	H, Me	H			166.50 [163.57]		3220			
II	H, Me	H	−12.25		168.79 [170.27]			1270		3400– 3000
III	H, Me	H		2.83		135.40 [137.74]		1270	1020 br	3300– 2900
I	Me, Me	H			168.07 [163.06]		3220 w			

continued

Table 3.4 *continued*

Type of compound	R^{ortho}	R^{para}	^{31}P NMR, δ (ppm)		^{13}C NMR, δ (ppm)		IR, ν (cm^{-1})			
			N–PO$_3$H$_2$	O–PO$_3$H$_2$	N–C=O[1]	N=C–O[1]	N–H	P=O	P–O–C	OH
II	Me, Me	H	−12.92		173.47 [162.99]			1205		3400– 2950
III	Me, Me	H		1.88, 2.56		137.80 [134.13]	3230 w	1265– 1250	1015, 1000	3100– 2900
I	Me, Me	Me			166.43 [163.06]					
II	Me, Me	Me	−13.19		173.22 [170.28]			1230		3250– 2950
III	Me, Me	Me		1.62, 2.51		138.70 [135.24]		1240, 1210	1025, 1005	3300– 3100
I	Pr^i, Pr^i	H			166.90[2] [169.16]		3250			
II	Pr^i, Pr^i	H	−12.65		167.60[2] [169.56]			1265		3400
III	Pr^i, Pr^i	H		2.02, 2.42	140.70[2] [137.95]		1225, 1190	1020, 995	3200– 2900	

The NMR spectra are registered in (CD$_3$)$_2$CO
1) Spectra are registered also in [(CD$_3$)$_2$SO]
2) Spectra were registered in CDCl$_3$
The IR spectra are registered in KBr

NMR was impossible because of poor solubility and low stability of the compounds in nonprotic solvents.

Though the signals in the region -12 to -13 ppm are attributed to both polyphosphates phosphorus atoms and the amidophosphates, it should be pointed out that the shape of the signals in these cases are considerably different. In many cases the shape of amidophosphates' signals is more simple and usually appeared as a singlet (Table 3.4).

The treatment of the N-phosphorylated benzanilides with a protic solvent such as methanol or ethanol or traces of very dilute hydrochloric acid, caused a decrease in intensity of the signals at -12.5 ppm and a simultaneous increase of the signal or two signals in the phosphate region between -2 and $+2$ ppm (Figure 3.3 'e' and 'f'). Hence one of the new low field signal(s) was assigned to the O-phosphorylated benzanilide (Table 3.4) [230]. This transformation in the phosphorus-31 NMR spectra matched precisely with the changes in the corresponding IR spectra, the last showed up two new intensive bands in the frequency region from 995 to 1025 cm^{-1} (Table 3.4) characteristic of the P–O–C bond. In the ^{13}C NMR spectra the N$\rightarrow$O migration of the dihydroxyphosphoryl group was reflected by two weak and broad signals at 135–142 ppm (N=$\underline{C}$–O–P) and the absence of the carbonyl carbon in the region 163–173 ppm [230] (Table 3.4). So the final products of benzanilides phosphorylation were corresponding O-phosphorylated derivatives (N-arylbenzimidoyl phosphoric acids) [227–230] (Scheme 45).

The signals of the same carbon atoms in different solvents [(CD$_3$)$_2$CO and (CD$_3$)$_2$SO] had different chemical shifts (Table 3.4). The ^{1}H NMR spectra of the phosphorylated benzanilides that contained the methyl substituents in the ortho position of the N-aryl fragment contained the two sets of signals attributed to these ortho methyl groups. The double sets or broadening of the corresponding signals in ^{31}P, ^{13}C and ^{1}H NMR and the two bands of P–O–C bands in IR spectra mentioned above strongly support the presence of E and Z stereoisomers in the phosphorylated benzanilides. Though the nonsubstituted product did not show distinctive evidence for stereoisomerism, this effect could not be excluded.

Detection of two E and Z tautomers of benzanilides was aided by the presence of substituents in the aromatic ring. This results in the slow rotation around the bonds Ar–N and C(O)–R [221], which is particularly evident for benz-2,6-diisopropylanilide.

The specific stereo properties of benz-2,6-diisopropylanilide structure resulted in the higher rate of transformation of its N-phosphorylated derivative into the corresponding O-phosphorylated product. The spectrometric characteristics of the O-phosphorylated benz-2,6-diisopropylanilide were fairly close to those of benz-2,6-dimethylanilides with differentiation of all carbon atoms of the isopropyl groups in ^{13}C NMR. In ^{1}H NMR spectra of the final product the four methyl groups had different chemical shifts, which indicates the presence of E and Z rotamers as is shown above for the methyl substituted N-phenylbenzimidoyl phosphoric acids [230].

The higher stability of the P–O bond compared with the P–N bond and the development of an extended conjugated system resulted in the irreversible N→O phosphorotropy between N-aryl-N-benzoylphosphoramidic acids and N-phenylbenzimidoyl phosphoric acids [229, 230]. According to molecular mechanics and semiempirical modelling, the coplanarity of the C=N bond with the aromatic systems was higher than of the coplanarity of the C=O bond in the N- and O-phosphorylated benzanilides.

UV irradiation of the solid sample of N-phosphorylated benzanilides under rigorous anhydrous conditions promoted the slow process of the N→O irreversible phosphorotropy which was monitored by the solid state ^{31}P and ^{13}C NMR spectroscopy [230]. These data illustrate for the mechanism of N→O phosphorotropy without participation of water or other protic solvents. Though the participation of the protic solvents cannot be ruled out, in the publication [230] the authors of the current book proposed a mechanism for the dihydroxyphosphoryl group N→O migration catalysed by a protic solvent which involves the formation of an intermediate four-member ring and stereo nonspecific elimination of the elements of a solvent. A similar mechanism of phosphorotropy is probable for phosphorylated benzoylamides and acetanilides as well (Scheme 86).

All attempts to induce a reverse O→N phosphorotropy in N-phenylbenzimidoyl phosphoric acids were unsuccessful [230].

Prolonged treatment of phosphorylated benzanilides with an excess of unreacted benzanilide at room temperature or 3–6 h refluxing in chloroform gave the products of condensation with elimination of phosphoric acid similar to phosphorylated acetanilides (Scheme 87) [224, 225] (Section 3.3.2). The structure of 1,2,3-triphenyl-3-benzoylamidines was determined by X-ray analysis.

Amide–imide rearrangement in benzanilides was described for the first time by Wislicenus and Goldschmidt [231] in 1900. Fifteen years later the acyl group migration in amides was studied by Mumm and co-workers [232] (Scheme 88).

In the following publications [233–237] it was named 'The Mumm rearrangement'. Authors of the publications [233, 234, 236] studied amides benzoylation and the consequent O→N migration of the functional group in detail (Scheme 89).

McCarthy, Hegarty and McCormack [238–240] developed the earlier expressed idea [233–236] concerning formation of a four-member cycle (Scheme 89) as an intermediate in the process of O→N migration of the acyl group which involves the prior Z(65%, p) = "rg9"E reorientation in imides (Scheme 90).

$$R = Alk, Ar; \quad R^1 = H, Alk \tag{86}$$

The kinetic study of the acyl group migration presented in the papers [241–243] determined that this process can be reversible and there was no significant direct influence of the substituents in aromatic rings upon the rate of the migration process somewhat similar to phosphorotropy in acetanilides (Section 3.3.2). The rate of the benzoyl group migration is five times lower than that of the acyl group [239] because of steric hindrance.

Another typical example of the 1–3 shift in substituted amides determined by Chapman [244–246] involved migration of the phenyl ring (Scheme 91).

The detailed reviews devoted to the Chapman rearrangement were published in 1965 [235], 1970 [247] and 1975 [248]. Electron-withdrawing substituents in the migrating groups promoted the rearrangement [249]. The alkyl substituents can migrate in amides as well [235, 250–254]. It is noteworthy that publications [244–249, 253, 255] were concerned with exclusively O→N irreversible shifts.

Nitrotropy was studied in detail by the authors of publications [256–260]. In many cases the NO_2 group migration ended up in the aromatic ring [256–265] (Scheme 92).

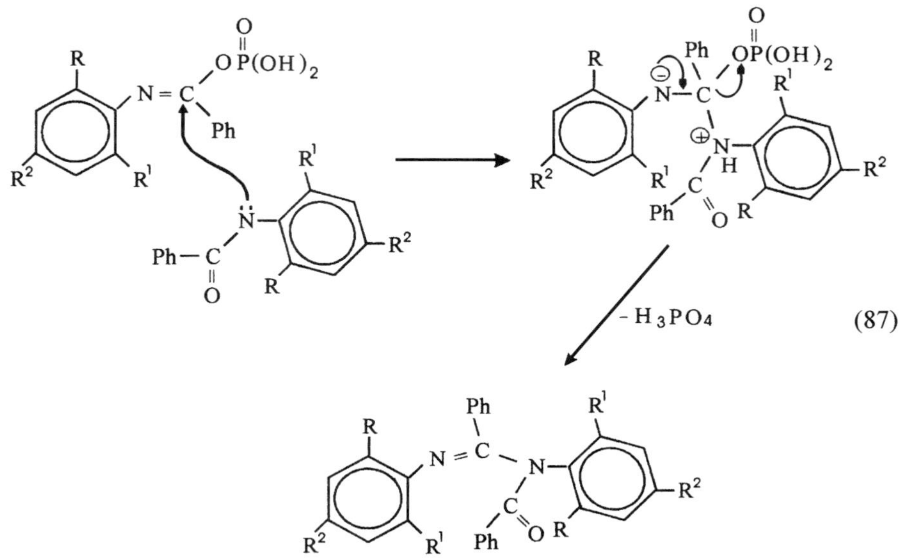

$$(87)$$

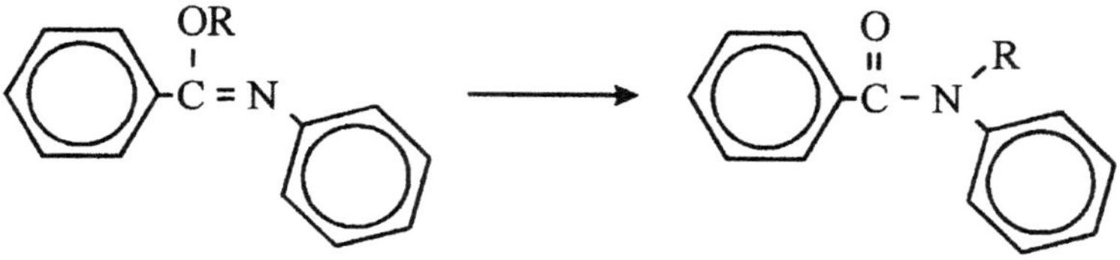

$$(88)$$

$$R = \underset{\underset{O}{\|}}{C} - Me \ , \quad \underset{\underset{O}{\|}}{C} - Ph$$

Chlorine [266] and trialkylsilanyl group [267] migrate in amides reversibly (Scheme 93).

There is a number of publications devoted to the synthesis of phosphorylated amides [268, 269] and thioamides [196, 270–276]. Common synthetic approaches to phosphorylated amides applied the reactions of amides with chlorophosphates [197, 270], imidoylchlorides with salts of dialkyl phosphates [273], phosphazaacyls with phenols [277], acetazides with trialkyl phosphite [278], acylamides with phosphorus pentachloride [278], and some others. In all cases the formation of full esters of phosphoric acids is described. The amide–imide equilibrium is characteristic of phosphorylated amides of thiobenzoic acid [270, 271], N-alkylthiobenzoyl amides [271, 275, 276] and thiobenzanilides [196, 276] (Scheme 94).

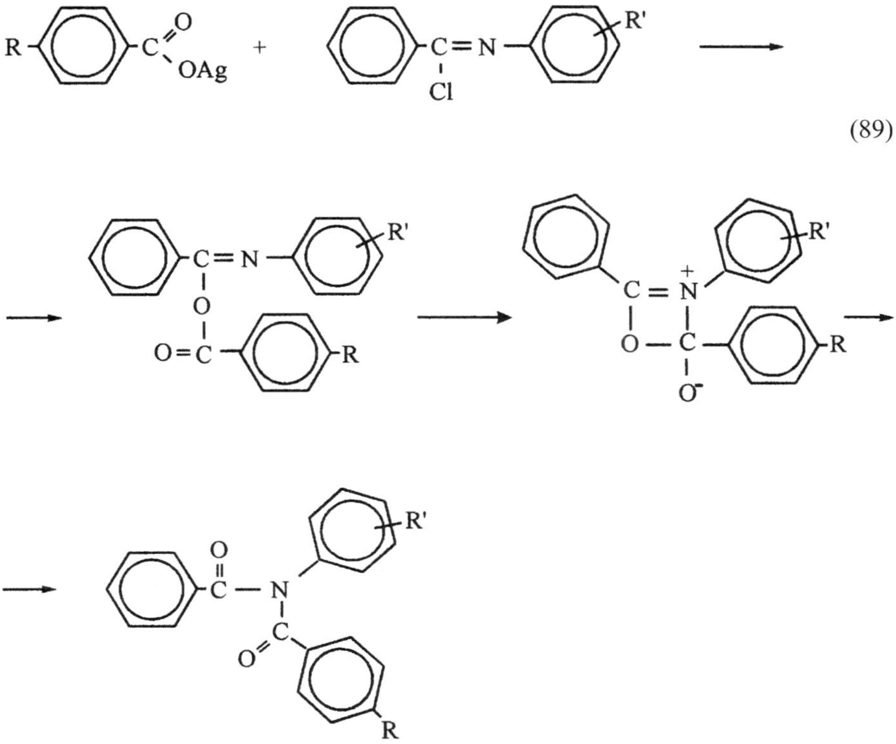

(89)

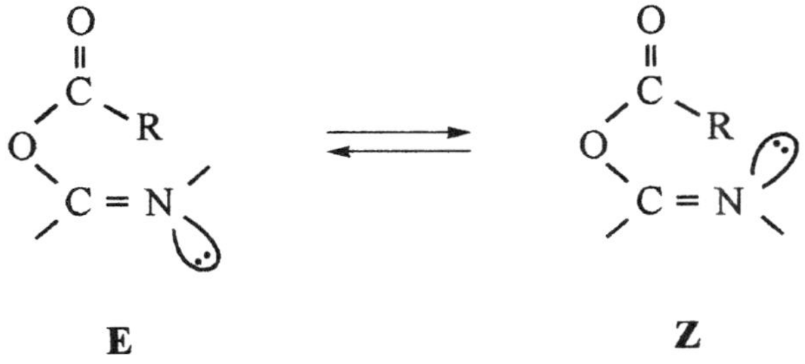

(90)

E   Z

The authors [276] proposed a cyclic dimeric form as an intermediate of the process (Scheme 95).

There are examples of unstable N-phosphorylated amides such as thiobenzamide [279] (Scheme 96). Probably decomposition takes place not because of the nature of the products but the particular method of their synthesis and isolation.

$$(91)$$

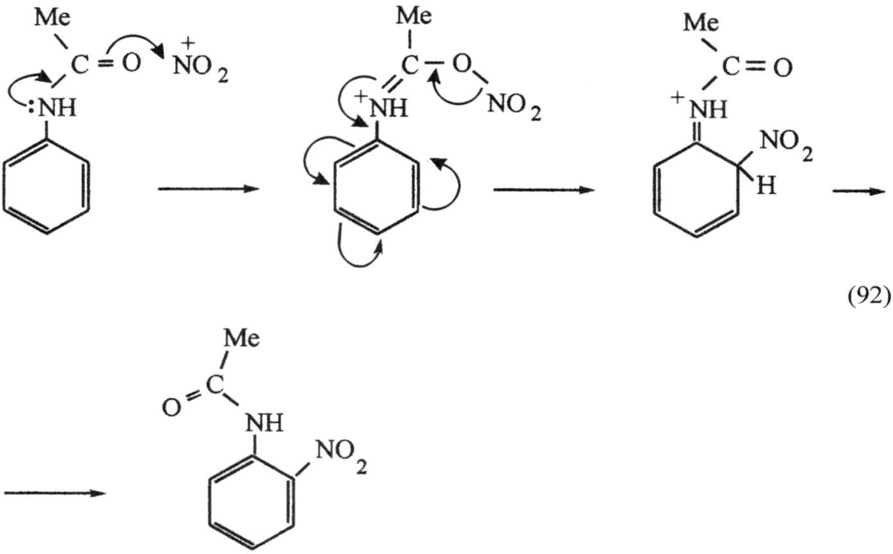

$$(92)$$

$$-\overset{O}{\underset{\|}{C}}-\overset{Z}{\underset{|}{N}}- \;\rightleftharpoons\; -\overset{O-Z}{\underset{|}{C}}=N- \tag{93}$$

$$Z = Cl, \; SiAlk_3$$

Zimin and co-authors [280] presented evidence for the irreversible nature of dialkoxyphosphoryl group S$\rightarrow$N shift under the influence of basic media. In publications [277, 278, 281–283] devoted to phosphorylated amides the authors did not discuss the effect of phosphorotropy. Nevertheless, Baraniak and Stec [284, 285] determined a reversible electrophile-assisted N$\rightarrow$O phosphorotropy in phosphorylated benzanilides (Scheme 97).

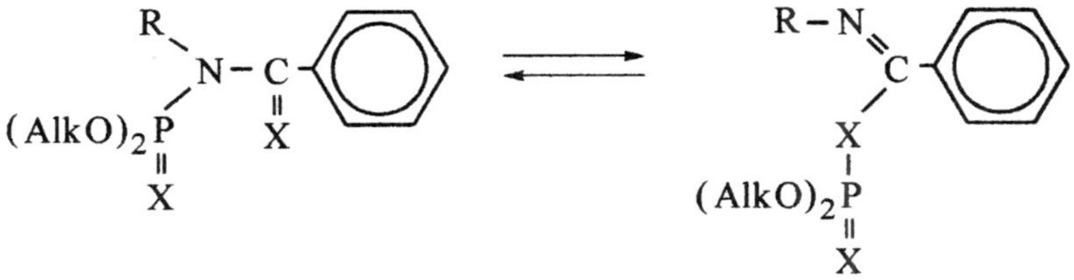

$$X = O, S \; ; \quad R = H, Me, Ph \tag{94}$$

$$X = O, S \; ; \quad R = H, Me, Ph$$

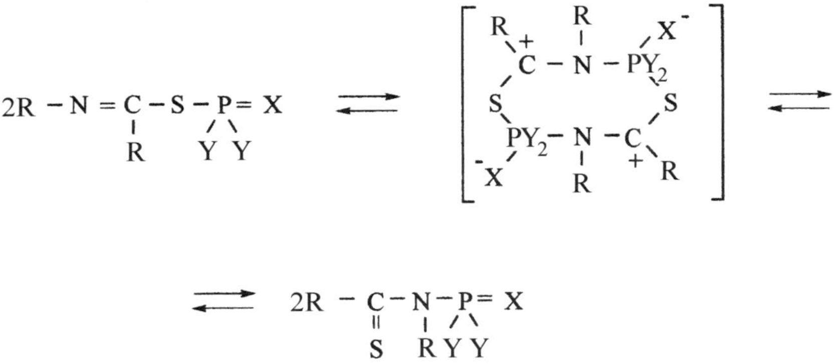

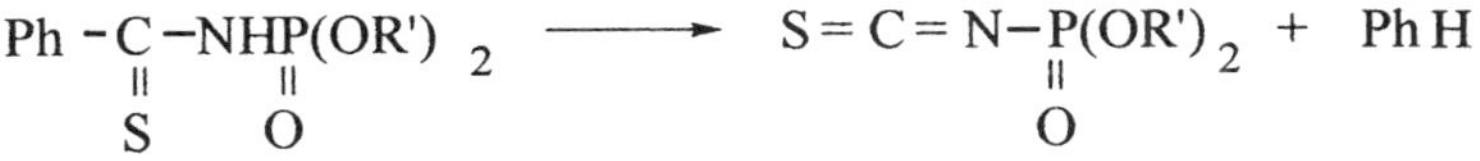

$$2R-N=C-S-P=X \; \rightleftharpoons \; \left[\begin{array}{c} \text{cyclic intermediate} \end{array} \right] \; \rightleftharpoons \tag{95}$$

$$\rightleftharpoons \; 2R-\underset{\underset{S}{\|}}{C}-\underset{\underset{R}{|}}{N}-\underset{Y\ Y}{\overset{/\backslash}{P}}=X$$

$$Ph-\underset{\underset{S}{\|}}{C}-NH\underset{\underset{O}{\|}}{P}(OR')_2 \; \longrightarrow \; S=C=N-\underset{\underset{O}{\|}}{P}(OR')_2 \; + \; PhH \tag{96}$$

$$(AlkO)_2\underset{\underset{O\ Ph}{\|\ |}}{P}-N-C\overset{O}{\diagup} \; \rightleftharpoons \; Ph-N=C\underset{Ph}{\overset{O\ \overset{O}{\|}}{\diagup}P(OAlk)_2} \tag{97}$$

It is highly probable that prototropy in amides promotes the process of phosphorotropy in the corresponding phosphorylated derivatives.

Migrations of substituents in amides plays a significant role in biochemical processes and pharmacology. For example preparation of triflouromethoxy-dibenzdiazepin which acts actively on the central nervous system involves the Chapman rearrangement [286] (Scheme 98).

$$(98)$$

A similar process was used for the synthesis of iodine containing analogues of tyroxine [287], an anti-inflammatory drug containing N-anthranilic acid fragment [288] and acridones [289]. A 1–3 shift of the acyl groups in the isothiobiotin fragments of enzyme·biotin·CO_2 complexes was proposed in the publication [290]. N-benzoyl-N-(dialkylphosphoryl) amide was found to act as a phosphorylating agent [291], which is not the case for the other phosphorylated amides [285].

3.4 REACTIONS WITH CYCLIC AMIDES AND THIOAMIDES

There have been studies of the reactions of phosphoric anhydride with five- and six-membered heterocyclic compounds with one or two amide and/or thioamide groups as part of the ring. These compounds are pyrrolidones (**21**), 2,4-dihydroxyimidazole (**22**), 4-hydroxy-2-mercaptoimidazole (**23**), 1,3-pyrimidine-2,4,6-triones (**24**), 1,3-pyrimidine-2-thione-4,6-dione (**25**).

These compounds possess some important characteristic features that makes their phosphorylation interesting. These include keto–enol and/or amide–imide tautomerism and high biological activity.

3.4.1 Reactions with five-membered cyclic amides

The authors of the current book have found that practically all five-membered cyclic amides have a low reactivity towards phosphoric anhydride. Many attempts to phosphorylate pyrrolidone (**21**), 4-phenylpyrrolidone and 2,4-dihydroxyimidazole (hydantoin) (**22**) with P_4O_{10} were unsuccessful. In all cases unreacted initial organic compounds were isolated. Such high stability of the cyclic compounds is probably due to low nucleophilicity of the amide nitrogen and the absence of prototropic processes. These reasons match nicely with a low order of C(O)–N bond in pyrrolidones determined by the authors of the publication [191]. The tautomeric transformations in hydantoin (**22**) are limited due to the aromatic system formed.

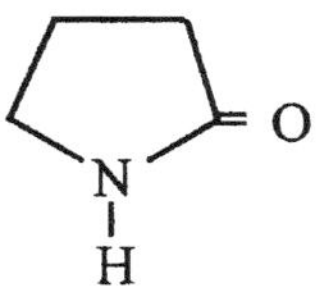

(21) 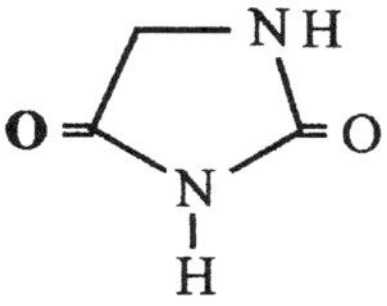 **(22)** 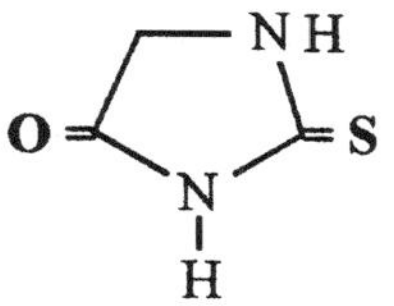 **(23)**

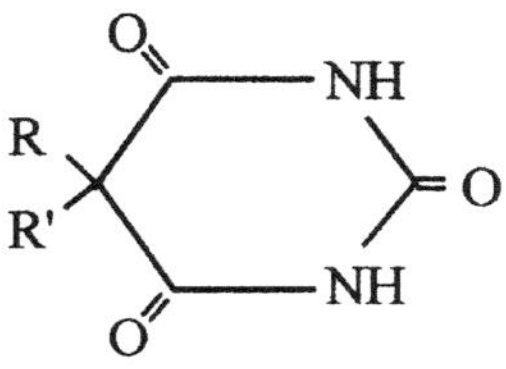

(24) 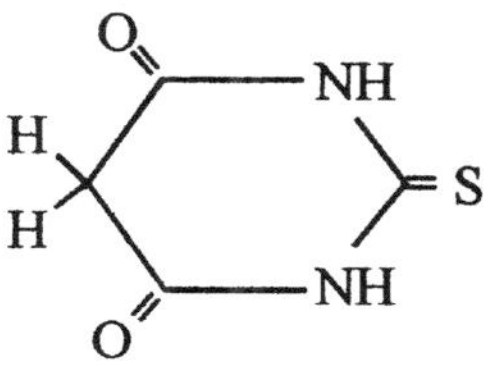 **(25)**

Phosphoric anhydride reacted with 4-hydroxy-2-mercaptoimidazole (thio-hydantoine) **(23)** that contains three nucleophilic centres (OH, NH, SH) [215]. The process was promoted by the presence of pyridine and resulted in S-phosphorylation and the formation of 2-(dihydroxyphosphorylthio)-4-hydroxyimidazole [165] (Scheme 99).

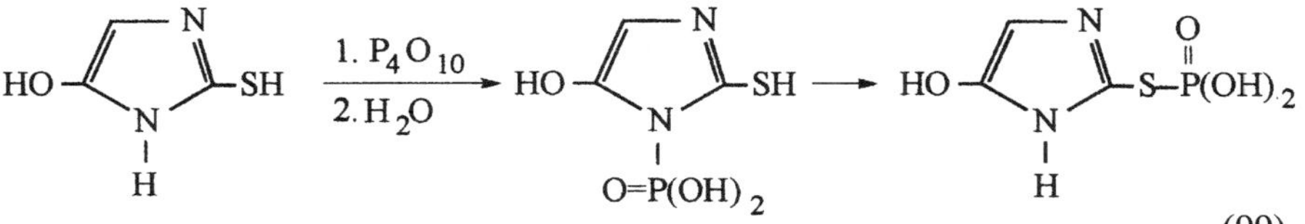

$$(99)$$

Nevertheless the ^{31}P NMR spectrum of the crystalline product immediately after isolation showed two signals with the shifts -13.64 and 7.31 ppm that were assigned to the N and S-phosphorylated forms, respectively. Fifteen minutes' storage of the product in methanol-d$_6$ resulted in the complete disappearance of the high field signal. This is a common feature with the reactions of acyclic amides with phosphoric anhydride (Section 3.3). The primary attack of thiohydantoine by P$_4$O$_{10}$ is directed towards the most nucleophilic nitrogen atom followed by consecutive hydrolysis of P–O–P bonds and eventually N$\rightarrow$S migration of the dihydroxyphosphoryl group.

3.4.2 Reactions with 1,3-pyrimidine-2,4,6-triones

1,3-Pyrimidine-2,4,6-triones can exist in three tautomeric forms 'a', 'b' and 'c' due to amide–imide equilibria [292–295] (Scheme 100). The form 'd' of the nonsubstituted compound (in position 5) arises from keto–enol tautomerization.

$$(100)$$

Molecular modelling indicated similar energies for the tautomeric forms and ^1H and ^{13}C NMR spectra supported mobile interconversions between the tautomers [296].

Molecular modelling of pyrimidinetriones [297] exhibited a more significant input of nitrogen atoms in the HOMO in comparison with oxygen. Actually this explains a high reactivity of pyrimidinetriones towards phosphoric anhydride that resulted in formation of the corresponding products of N-phosphorylation [165, 215, 298]. The substituents at position 5 of the heterocycles influenced reactivity of the compounds. The unsubstituted pyrimidinetrione reacted with phosphoric anhydride in the presence of a catalytic amount of pyridine (yield 46%) (Scheme 101). Probably pyrophosphate was formed as the result of phosphorylation of monosubstituted phosphate by either unreacted phosphoric anhydride or polyphosphoric acid which was formed in the process [79].

$$(101)$$

The substituents at the C5 atom (R=R^1=C$_2$H$_5$; R=C$_2$H$_5$, R^1=C$_6$H$_5$) of pyrimidinetriones increased the higher nucleophilicity of nitrogen atoms [299] which resulted in their higher reactivity in the reactions with phosphoric anhydride (Scheme 102).

$$\text{(102)}$$

R,R$'$ = Et, Ph

After 60–90 min stable products of N-phosphorylation were isolated in yields of about 70–90% [298]. ^{13}C NMR spectra exhibited the isomeric homogeneity for each product.

Possible participation of pyrimidinetriones β-diketone fragment in the reaction with phosphoric anhydride was modelled by the reaction of the latter with 5,5-dimethylcyclohexyl-1,5-dione (dimedone). The reaction took 12–18 h and gave 5,5-dimethyl-1-dihydroxyphosphoryloxycyclohexene-3-one with the yield 56% [300] (Scheme 103).

$$\text{(103)}$$

This experiment indicates that for pyrimidinetriones, N-phosphorylation is much faster than O-phosphorylation.

3.4.3 Reactions with 1,3-pyrimidine-2-thione-4,6-dione

It was shown earlier (Section 3.4.1) that the reactions of phosphoric anhydride with five-membered heterocycles proceed more readily when a thione group is present. In this respect a similar comparison of six-membered rings with and without the thione group is of particular interest.

Molecular modelling of 1,3-pyrimidine-2-thione-4,6-dione was well supported by the experimental data.

Reaction of phosphoric anhydride with 1,3-pyrimidine-2-thione-4,6-dione in chloroform proceeded under conditions milder than those of 1,3-pyrimidine-2,4,6-trione (Section 3.4.2) and gave a phosphorylated product with the yield 70% [215] (Scheme 104).

$$(104)$$

^{31}P NMR spectra showed the reaction to be regioselective, the attack by phosphoric anhydride occurring at nitrogen only. Treatment of the crystalline product with acetone resulted in isolation of N-dihydroxyphosphoryl-1,3-pyrimidine-4,6-dione-2-thione. Under the action of a protic organic solvent the dihydroxyphosphoryl group migrated from the N to S atom [215] (Scheme 105).

$$(105)$$

2-S-Dihydroxyphosphorylthio-1,3-pyrimidine-4,6-dione underwent a reversible phosphothiol–phosphothione rearrangement registered by ^{31}P NMR. Phosphothiol–phosphothione equilibrium is of high significance for the sulphide sensitization in silver halide photographic materials [301]. Similar types of tautomeric transformations were studied by several research groups [302, 303] and in particular by Michalski and co-workers [304–310].

In ^{13}C NMR spectra of the product atoms C5 appeared as two signals (82.8 and 85.5 ppm) which can be assigned to enolisation and the formation of an endocyclic double bond. Stabilization of such a type of barbiturates structures was established by Balakina and Zuev [299].

So, the thione group in five- and six-membered cyclic amides have a significant influence upon the reactions of these heterocycles with phosphoric anhydride and the nature of products.

3.4.4 Reactions with hydrazide of benzoic acid

The reaction of phosphoric anhydride with the hydrazide of benzoic acid was vigorous and resulted in phosphorylation of the terminal nitrogen atom [311] (Scheme 64). ^{31}P NMR monitoring showed a similar pattern of the reaction as shown above for acyclic amides (Section 3.3). The main difference in these processes was the relative stability of the polyphosphate intermediates probably due to the strong hydrogen bonding. Mild treatment of the polyphosphates with water resulted in P–O–P bond hydrolysis with the formation of N-benzoylhydrazido phosphate. Addition of catalytic quantities of hydrochloric acid initiated a slow reversible 1–4 (N→O) shift of the dihydroxyphosphoryl group [311] (Scheme 106).

$$(106)$$

It is likely that the relative instability of the O-phosphorylated form is determined by the attachment of the phosphoryl group to the intermediate member of the conjugated system which weakens the π-electronic system [312–315].

3.5 REACTIONS WITH AZOLES

One of the most characteristic features of azoles is 1–3 prototropy (Scheme 107) which has been studied in depth by means of experimental and theoretical methods [316–336].

$$X = CH, N \qquad (107)$$

Probably azoles can exhibit another 1–2 prototropy (Scheme 108) which gives a quinonoid structure.

$$(108)$$

By means of dipole moments in combination with quantum chemical calculations (INDO) [335] it was shown that the 1–H form of benzotriazole dominates entirely over the 2–H form. Recently acylotropy in triazoles was described in the paper [336].

From a practical point of view azoles and particularly benzazoles are very important compounds. Benzimiazole and its phosphorylated derivatives are used in ophthalmology [337, 338], they are active anti-tumour, anti-ulcer and cardiotonic drugs [339–342] and possess others types of activity [343, 344]. Phosphorylated benzimidazoles are very active growth stimulants of agricultural plants [345–349a]. Azoles can be used as a component of explosives [350], and as mediators in organic synthesis [321, 351]. They are also widely used as stabilizers and crystal growth modifiers in silver halide photographic emulsions [352–354].

In our experiments unsubstituted imidazole was phosphorylated by phosphoric anhydride to give diimidazolyl phosphoric acid in a moderate yield (48%) (Scheme 109). The reaction was accompanied by the partial decomposition of the heterocyclic ring.

$$(109)$$

Under mild conditions ($CHCl_3$ media, $40°C$) 1,2,3-triazole did not react with phosphoric anhydride but under more forcing conditions (long refluxing in $CHCl_3$ with an excess of P_4O_{10}) the triazole cycle decomposed completely.

Benzimidazole, benzotriazole and their substituted in positions 5 and/or 6 derivatives were readily phosphorylated by phosphoric anhydride [215, 227, 354–357]. The products of the reactions were corresponding 1-(benzimidazolyl) or 1-(benzotriazolyl) phosphoric and pyrophosphoric acids (Scheme 110). The addition of a catalytic quantity of tetrahydrofuran in the reaction mixtures directed the reaction towards the prevailing formation of amidopyrophosphates [356, 357].

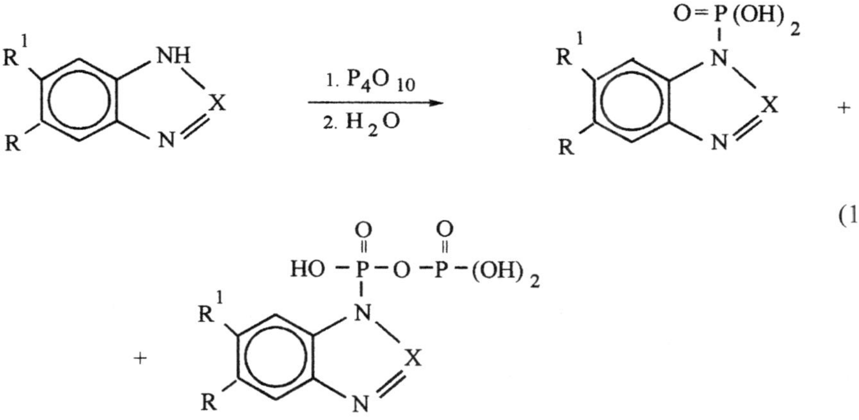

$$\text{(110)}$$

$$R, R^1 = H, Me, NO_2$$

5(6)-Methylbenzotriazole gave only the phosphoramidic acid and 5,6-dimethylbenzotriazole produced entirely the amidopyrophosphoric acid [355, 356]. The location of dihydroxyphosphoryl group in 1-(5-methylbenzotriazolyl) phosphoric acid was determined by ^1H and ^{13}C NMR spectrometry [356, 357].

The formation of amidopyrophosphates by phosphorylation of amido-phosphates with phosphoric anhydride or polyphosphoric acid was possible according to Scheme 111.

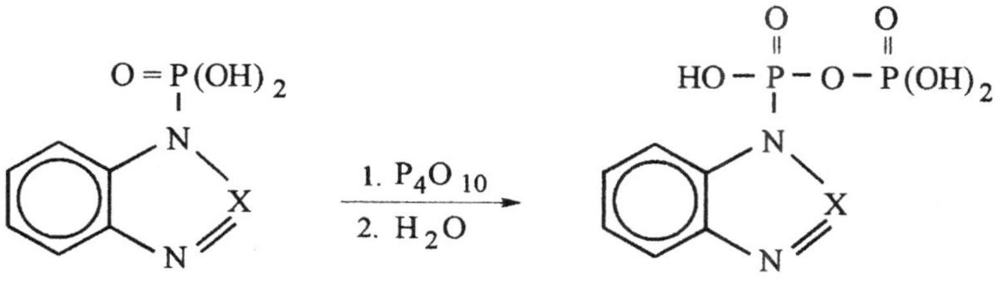

$$\text{(111)}$$

$$X = CH, N$$

The patterns of ^{1}H and ^{13}C NMR spectra of 1-(5,6-dimethylbenzotriazolyl) pyrophosphoric acid were surprising with respect to the magnetic equivalence of the corresponding methyl groups nuclei. The equivalence was attributed to a facile phosphorotropy between the two nitrogen atoms (N1 and N3) (Scheme 112) in polar solvents (acetone-d$_6$, DMSO-d$_6$, methanol-d$_4$) [356].

$$(112)$$

$$\textcircled{P} = \underset{\underset{O}{\|}}{P}(OH)_2, \quad \underset{\underset{O}{\|}}{\overset{\overset{OH}{|}}{P}}OMe, \quad \underset{\underset{O}{\|}}{\overset{\overset{OH}{|}}{P}}O\underset{\underset{O}{\|}}{P}(OH)_2$$

Mild hydrolysis of the P–O–P bonds in 1-(5,6-dimethylbenzotriazolyl) pyrophosphoric acid gave the corresponding amidophosphoric acid which, along with its partially esterified derivative, also exhibited 1–3 phosphorotropy [356]. The ^{13}C NMR spectra which were determined on pyridine-d$_5$ solutions showed magnetic nonequivalence of the methyl groups. However the generation of pyridinium salt of the pyrophosphoric acid retards phosphorotropy. The alternative localization of the dihydrophosphoric group on the N2 atom as a cause of the equivalence was rejected since the formation of the o-quinonoid structure should cause a distinctive electronic structure [302] which was not evident spectroscopically [356, 357].

Phosphorylation of the N–H bond in azoles has been achieved by various methods using a wide variety of phosphorylating agents. Usually these are phosphorus trichloride and alkylhalogen phosphates [358–360], phosphorus oxychloroxide [361–364], esters of phosphonic and phosphorous acids [345, 349, 365–368] (Scheme 113) and some other compounds [369].

$$(113)$$

$$n = 1\text{-}3; \quad X = O, S; \quad R = OAr, NHAr$$

Phosphorylated imidazoles have been prepared by the Atterton–Todd reaction [370–372] (Scheme 114) and also by insertion into the pyrophosphate bond [371–373] (Scheme 115).

$$\text{imidazole–NH} + \text{H}\overset{\text{O}}{\text{P}}(\text{OR})_2 \xrightarrow{\text{CCl}_4} \text{imidazole–N–}\overset{\text{O}}{\text{P}}(\text{OR})_2 \qquad (114)$$

$$(115)$$

Phosphorylation of imidazoles was achieved also by the use of monoalkyl-phosphate in a redox process [374] (Scheme 116).

There are several examples of specific processes of the N–H bond for the phosphorylation of the N–H bond in imidazoles and their derivatives [374, 375].

3.6 REACTIONS WITH O-DIAMINOBENZENES

The reactions of phosphoric anhydride with N-acetyl and N-benzoyl-o-diaminobenzenes are complicated by the presence of two nucleophilic groups. N-benzoyl-o-diaminobenzene exhibited a higher reactivity towards phosphoric anhydride than N-acetyl-o-diaminobenzene. Five and twelve hours refluxing in chloroform, respectively gave precipitates that did not contain phosphorus. The products were shown to be 2-phenylbenzimidazole and 2-methylbenzimi-dazole [215, 376] (Scheme 117). Evaporation of the solvent and analysis of solid residues gave respectively 1-(2-phenylbenzimidazolyl) phosphoric acid plus 1-(2-phenylbenzimidazolyl) pyrophosphoric acid and 1-(2-methylbenz-imidazolyl) phosphoric acid plus 1-(2-methylbenzimidazolyl) pyrophosphoric acid [215, 376] (Scheme 117). Amidopyrophosphoric acids were highly hydroscopic compounds.

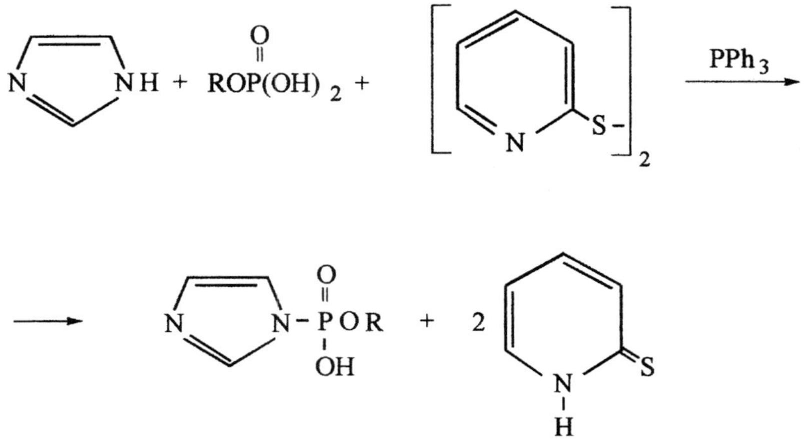

$$(116)$$

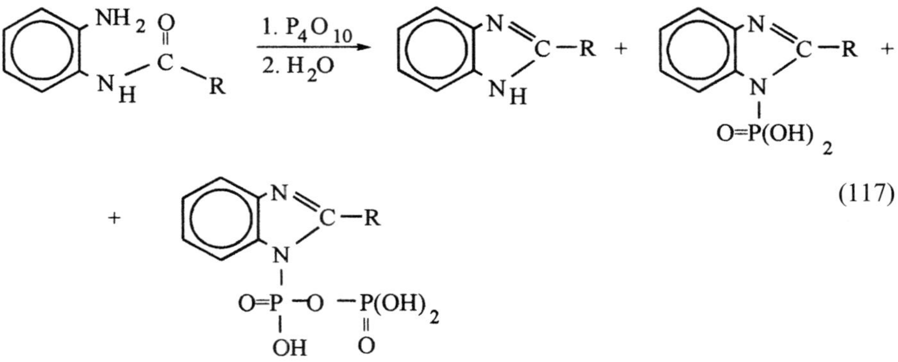

$$(117)$$

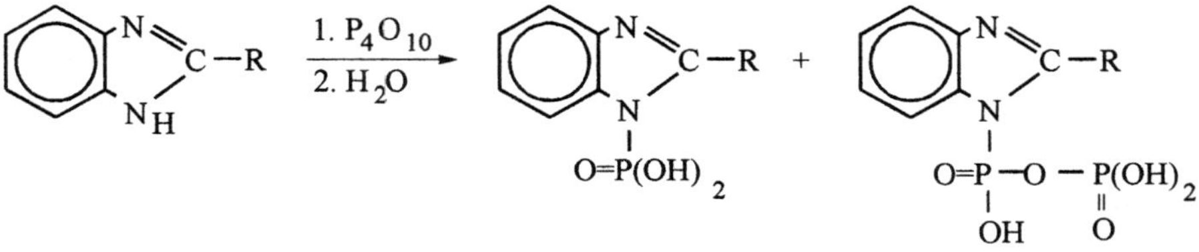

R = Me, Ph

$$(118)$$

The proposed reaction pathway involved initial phosphorylation of the more nucleophilic amino group. The following intramolecular dehydration gave 2-substituted benzimidazoles. This intramolecular condensation is somewhat similar to the intermolecular condensation between phosphorylated acetanilides and benzanilides with unreacted amide (Sections 3.3.2 and 3.3.3).

Phosphorylated 2-methyl and 2-phenyl benzimidazoles were synthesized independently in the reactions of the corresponding 2-substituted benzimidazoles with phosphoric anhydride (Scheme 118) according to the method used for nonsubstituted benzimidazoles (Section 3.5).

The process based upon P$_4$O$_{10}$ presented above is another new synthetic approach to substituted benzimidazoles in addition to the well-known and recently proposed methods that employed predominantly the reactions of o-diaminobenzene with acetic or benzoic acid, their esters, amides or nitriles under the action of polyphosphoric acids, polyphosphoric esters and other dehydrating-condensing agents [377–380]. The method described in publication [377] is discussed below (Section 6).

4 REACTIONS OF PHOSPHORIC ANHYDRIDE WITH PHOSPHORUS-CONTAINING COMPOUNDS

Phosphoric anhydride reacts with phosphoryl trichloride at 200°C to give a complicated mixture of products [381, 382] the major constituents being pyrophosphoryl chloride and some of the cyclic trimetaphosphoryl chlorides [383]. The former provides a useful route to phosphoric acid esters of the type ROP(O)Cl$_2$ [384–387] (Scheme 119). The nature and yields of products formed depends upon the ratio of the initial reagents, time of contact and temperature.

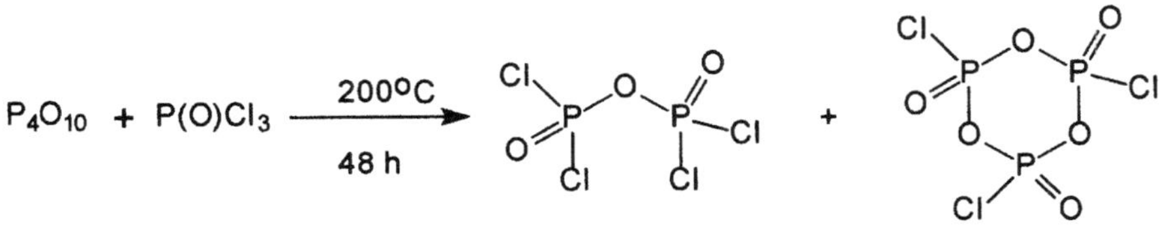

$$P_4O_{10} + P(O)Cl_3 \xrightarrow[48\ h]{200°C}$$

(119)

Phosphorus-31 NMR spectrometry indicated the presence of tetrachloropyrophosphates and chloropolyphosphoric structures including partial ring opening of the P$_4$O$_{10}$ cage structure among the products of the process [382]. The following correlation between P (ppm) and the phosphorus groups seems to be appropriate:

$$+2 \text{ POCl}_3, \quad -7 \text{ to } -11 \ -\text{OPOCl}_2, \quad -28 \ (-\text{O})_2\text{POCl}, \quad -50 \pm 3 \ (-\text{O}_3)\text{PO}.$$
$$\qquad\qquad\qquad\quad \text{ends} \qquad\qquad\qquad \text{middles} \qquad\qquad\qquad \text{branched}$$

Esters of phosphonic acids were produced in the reaction of phosphoric anhydride with various amounts of trimethyl and triethyl phosphates [388, 389]. The authors considered the equilibrium polymeric species of different structures. Unlike trialkyl phosphates the reaction of triphenyl phosphate with phosphoric anhydride (molar ratio 2:1) according to Berger [390] leads to the formation of branched phenyl polyphosphate with no equilibrium effects.

Schwarzmann and Van Wazer [155] carried out a ^{31}P NMR study of the products of the reaction $P_4O_{10}+P(O)(OAlk)_3$. On the basis of comparative analysis of the data presented in publications [147, 390a] they determined the following chemical shifts for the resultant polyphosphates (Alk=Me, Et): -1 (ortho), -13 (middle), -28 (terminal) and -41 (branched) ppm.

The result of the reaction of phosphoric anhydride with trialkylphosphates at 90°C depended upon the reagents' ratio [391, 392]. The molar ratio $P_4O_{10}:(EtO)_3P=O$ 1:4 favoured the formation of hexaethylenetetrapolyphosphate (Scheme 120) and the ratio 1:8 resulted in isolation of tetraethylpyrophosphate (Scheme 121). The formation of tetraethylpyrophosphoric acid was proved by a model synthesis from silver pyrophosphate and ethyl iodide [392] (Scheme 122).

$$P_4O_{10} + 4(EtO)_3\,P=O \xrightarrow{\ 90^{\mathrm{o}}C\ } 2(EtO)_2\,\overset{\overset{O}{\|}}{P}-O-\overset{\overset{O}{\|}}{\underset{\underset{OEt}{|}}{P}}-O-\overset{\overset{O}{\|}}{\underset{\underset{OEt}{|}}{P}}-O-\overset{\overset{O}{\|}}{P}\,(OEt)_2 \qquad (120)$$

$$P_4O_{10} + (EtO)_3\,P=O \xrightarrow{\ 90^{\mathrm{o}}C\ } \qquad (EtO)_2\,\overset{\overset{O}{\|}}{P}-O-\overset{\overset{O}{\|}}{P}(OEt)_2 \qquad (121)$$

$$(AgO)_2\,\overset{}{\underset{\underset{O}{\|}}{P}}-O-\overset{}{\underset{\underset{O}{\|}}{P}}(OAg)_2 \xrightarrow[{-AgI}]{+EtI} \qquad\qquad\qquad\qquad\qquad (122)$$

The publication [157] presents a synthetic method for ethyl and o-cresyl polyphosphates from corresponding esters of phosphoric acid and P_4O_{10} but tri (p-nitrophenyl) phosphate did not react with phosphoric anhydride under similar conditions. Comparison of the ^{31}P NMR data determined for alkyl (aryl) substituted polyphosphates [76, 155, 156, 157] exposes a distinctive magnetic sensitivity of phosphorus to the nature of substituents.

Phosphoric anhydride reacts with amides of phosphorous and phosphinous acids [393] to give mixed amides of tri and pentavalent phosphorus (Scheme 123).

$$\text{R}_2\text{P--NEt}_2 + \text{P}_4\text{O}_{10} \longrightarrow \text{R}_2\text{P--O--}\overset{\overset{\text{NEt}_2}{|}}{\underset{\underset{\text{O}}{\|}}{\text{P}}}\text{--O--}\overset{\overset{\text{NEt}_2}{|}}{\underset{\underset{\text{O}}{\|}}{\text{P}}}\text{--OPR}_2 \qquad (123)$$

Similar pathways, with the formation of the P^{III}–O–P^{V} bonds, took place in the reactions of phosphoric anhydride with di and triamides of trivalent phosphorus [393]. The reaction mixture was distilled at high temperature which caused a partial decomposition of the products.

The mechanism of mixed anhydride formation is presented by the authors [393] as a nucleophilic attack of anhydride phosphorus by an unshared pair of electrons on nitrogen followed by heterolytic decomposition of the P^{III}–N bond. Schwarzmann and Van Wazer [394] studied the interaction of phosphoric anhydride with hexamethylphosphoryltriamide. The reaction was very slow, even at a high temperature (80°C or 160°C) and the equilibrium mixture contained several sets of chemical shifts in the ^{31}P NMR spectra. The following correlation between δ_{P} and the phosphorus environments seem to be appropriate.

$$23\,(\text{Me}_2\text{N})_3\text{P} = \text{O}, \quad 11.5 - \text{O}\,(\text{Me}_2\text{N})_2\,\text{P} = \text{O},$$
$$\text{ends}$$

$$-12\,(-\text{O})_2\,(\text{Me}_2\text{N})\,\text{P} = \text{O}, \quad -40\,(-\text{O})_3\,\,\text{P} = \text{O}\,\,.$$
$$\text{middles} \qquad\qquad\qquad \text{branched}$$

Phosphorus oxides (III, V) can be prepared in the reaction of phosphoric anhydride with red phosphorus [39]. It is noteworthy that phosphoric thioanhydride reacts with phosphoric anhydride under relatively mild conditions to produce phosphorus oxysulphides, one example is illustrated by Scheme 124.

$$4\,\text{P}_4\text{S}_{10} + 6\,\text{P}_4\text{O}_{10} \longrightarrow 10\,\text{P}_4\text{O}_6\text{S}_4 \qquad (124)$$

The mixed phosphorus oxysulphides are characterized by the 'cage-like' P$_4$ tricyclic structure.

The widespread use of phosphoric thioanhydride in the lubricant and extraction industries and with respect to biological applications arise from its reactivity towards the other phosphorus and heavy metals halides [395, 396] (Schemes 125–127).

$$\text{P}_4\text{S}_{10} + 6\,\text{PCl}_5 \longrightarrow 10\,\text{PSCl}_3 \qquad (125)$$

$$\text{P}_4\text{S}_{10} + 2\,\text{SbBr}_3 \longrightarrow 2\,\text{PSBr}_3 + 2\,\text{Sb}{-}\text{S}{-}\overset{\overset{\displaystyle\diagup\text{S}\diagdown}{}}{\underset{\underset{\diagdown\text{S}\diagup}{}}{\text{P}}}{=}\text{S} \qquad (126)$$

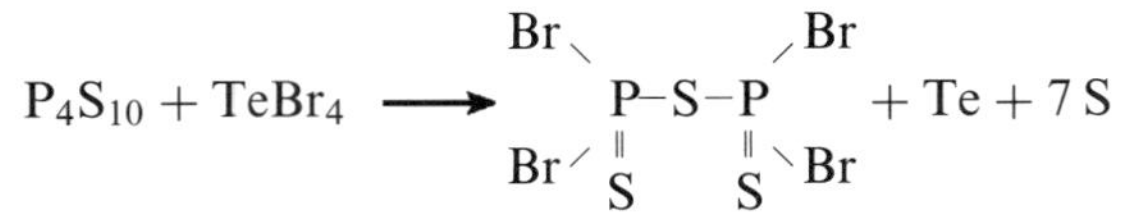

$$P_4S_{10} + TeBr_4 \longrightarrow \quad \underset{\substack{Br}}{\overset{\substack{Br}}{P\text{-}S\text{-}P}} \quad + Te + 7\,S \qquad (127)$$

5 REACTIONS OF PHOSPHORIC ANHYDRIDE WITH ARENES AND C–H NUCLEOPHILES

The majority of studies of the phosphorylation of aromatic compounds were carried out by Lecher and co-workers. They developed a unique method for the insertion of the phosphoryl group in an aromatic ring: 10 h heating of the reaction mixture, composed of phosphoric anhydride and the aromatic compound, in a sealed glass tube [397] or an autoclave [398] up to the temperature 250–325°C followed by treatment with hot water. This method was applied to benzene, chlorobenzene, o-xylene, naphthalene [397, 398], toluene, pyrene, chrysene and phenanthrene [398]. The most important outcome of these reactions is the formation of the new P–C bonds. According to the authors [397, 399] the high temperature of the process is required for either homolytic or heterolytic dissociation of P_4O_{10} (Scheme 128).

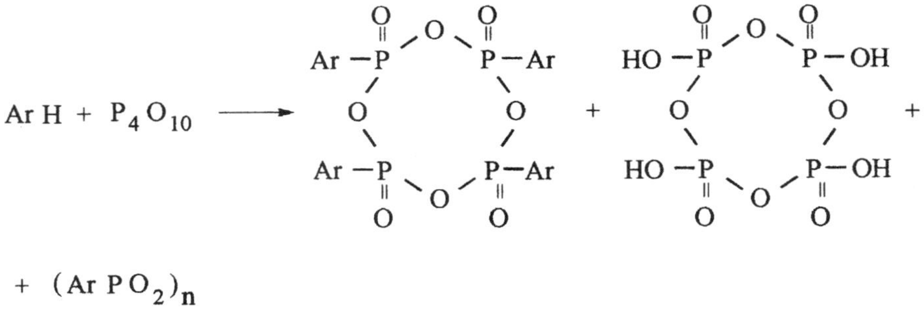

$$Ar\,H + P_4O_{10} \longrightarrow$$

$$+ \ (Ar\,PO_2)_n$$

$$(128)$$

An ionic mechanism is the most favoured one [397]. Both mechanisms could involve thermolysis of phosphoric anhydride to produce a reactive phosphorus species which caused electrophilic substitution of aromatic compounds similar to that caused by the nitronium cation $^+NO_2$ [397]. The primary step of the interaction was envisaged as the addition of one aromatic molecule to a partially opened form of phosphoric anhydride [397–400]. Chlorobenzene, o-xylene and naphthalene had a tendency to break down the second P–O–P bridge bond of the bicyclic polyphosphoric fragment thus giving respectively cyclotetrametaphosphoric and/or polymeric structures $(ArPO_2)_n$ [397, 400–402] (Scheme 128).

The isolation of the reaction products was based on their different solubility in the reaction mixtures. The products of primary interaction were insoluble in the aromatic compounds whereas the secondary products were highly soluble in the same solvents. The authors [397, 400] report a one-pot phosphorylation by phosphoric anhydride and consecutive chlorination by phosphorus pentachloride of the aromatic compounds with the formation of $ArPOCl_2$.

It is relevant to mention here that the properties of phosphoric anhydride in the reactions with aromatic compounds differ from those of P_4S_{10} [399]. Phosphoric anhydride produced the new P–C bonds under forcing conditions whereas tetraphosphorus decasulphide reacts with aromatic compounds under milder conditions to give the products of thiophosphorylation.

For example the Friedel–Crafts reaction was extended to the thiophosphorylation of aromatic systems by using phosphoric thioanhydride in the presence of aluminium chloride which also leads to new P–C bonds [1] (Scheme 129).

$$8\,C_6H_6 + P_4S_{10} \xrightarrow{\text{AlCl}_3} 4\,(C_6H_5)_2P{\overset{\displaystyle S}{\underset{\displaystyle SH}{<}}} + 2\,H_2S \qquad (129)$$

Another possible approach to the P–C bond formation was based on the prolonged heating of cyclohexane with a large excess of P_4S_{10} (Scheme 130).

$$4\,\langle H \rangle + P_4S_{10} \longrightarrow 2\,\langle H \rangle{-}P{\overset{S}{\underset{S}{<}}_{S}}P{-}\langle H \rangle + 2\,H_2S \qquad (130)$$

The most recent publication [402a] describes the possibility of P–C bond formation in the reaction of P_4O_{10} with HCN or KCN promoted by an organic base.

6 REACTIONS OF PHOSPHORIC ANHYDRIDE WITH SILICON-CONTAINING COMPOUNDS

The stimulating point for numerous studies devoted to reactions of phosphoric anhydride with a wide range of organosilicon compounds was Sauer's publication of 1944 [403] that presented the preparation of tris(trimethylsilyl) phosphoric acid. The product was distilled from the boiling reaction mixture, phosphoric anhydride–trimethylsilanol (Scheme 131).

$$\text{Me}_3\text{SiOH} \xrightarrow{\text{P}_4\text{O}_{10}, \text{ distillation}} (\text{Me}_3\text{SiO})_3\text{P=O} \qquad (131)$$

Two years later Turner and co-workers [404] found it impossible to dry silicon oil with phosphoric anhydride because a reaction occurred (Scheme 132).

$$8 - \underset{|}{\overset{|}{\text{Si}}} - \text{OH} \; \underset{\text{H}_2\text{O}}{\overset{\text{P}_4\text{O}_{10}}{\rightleftharpoons}} \; 2 - \underset{|}{\overset{|}{\text{Si}}} - \text{O} - \overset{\text{O}}{\underset{\text{O}}{\overset{\|}{\text{P}}}} - \text{O} - \overset{\text{O}}{\underset{\text{O}}{\overset{\|}{\text{P}}}} - \text{O} - \underset{|}{\overset{|}{\text{Si}}} - \qquad (132)$$

On the basis of experimental data Orlov and Voronkov [405] proposed a pathway for the reaction of phosphoric anhydride with trialkyl (aryl) hydroxysilanes (Scheme 133).

(133)

This interaction proceeded up to the complete decomposition of phosphoric anhydride and the formation of trisilyl phosphates.

The authors [405, 406] did not rule out another competing process that involves the high tendency of hydroxysilanes to undergo intermolecular dehydration and formation of disiloxanes. The latter compounds could undergo consequent reactions with phosphoric anhydride to give tris (trialkylsilyl) phosphoric acid (Schemes 134, 135) similarly to the mechanism presented above according to Scheme 133.

$$6\,R_3SiOSiR_3 \;+\; P_4O_{10} \longrightarrow 4\,(R_3SiO)_3P=O \tag{134}$$

$$n\,R_3SiOSiR_3 \;+\; nP_4O_{10} \longrightarrow 4 \left(\underset{\underset{R_3SiO}{\displaystyle|}}{\overset{\overset{\displaystyle O}{\|}}{P}} - O \right)_n \tag{135}$$

$$R = Me,\ Et,\ Pr,\ Bu$$

In fact compounds with Si–O–Si bonds have been found to be reactive towards phosphoric anhydride [405–407]. The rate of reaction decreases with the length of the alkyl chain (R). The products were isolated by distillation *in vacuo* and the yields of the products were determined by their basic hydrolysis [406] and decreased with an increase in the size of R. For tris(trimethylsilyl) phosphoric acid the yield 80% was determined according to this method.

The data presented [406] were reinvestigated by Mileshkevich and Karlin [408]; their yields of tris(trimethylsilyl) phosphoric acid did not exceed 25%. Moreover by high vacuum distillation pyrometaphosphates of the common formula presented below were isolated [408] in a total yield of 52% (Scheme 136).

$$(5-n)\,Me_3Si-O-SiMe_3 \;+\; P_4O_{10} \longrightarrow (2-n)(Me_3SiO)_3P=O \;+$$

$$+\; (Me_3SiO)_2 \underset{\underset{Me_3SiO}{\displaystyle|}}{\overset{\overset{\displaystyle O}{\|}}{P}} - O \left(\overset{\overset{\displaystyle O}{\|}}{P} - O \right)_n \overset{\overset{\displaystyle O}{\|}}{P} (OSiMe_3)_2 \tag{136}$$

$$n = 0,\ 1,\ 2$$

The reaction of phosphoric anhydride with hexamethyldisiloxane was carried out [408] under prolonged reflux and the products were isolated by vacuum distillation.

Okamoto [102] produced trimethylsilyl polyphosphates from hexamethyl disiloxane and phosphoric anhydride that under consequent treatment with an alcohol and water underwent desilylation with the formation of monoalkyl (aryl) phosphoric acids as presented in Scheme 137. This process is remarkable because of the selective alkylation of phosphoric anhydride by the alcohol leading to the formation of monoalkyl phosphoric acid and regeneration of hexamethyldisiloxane.

$$Me_3Si-O-SiMe_3 \xrightarrow{\;P_4O_{10}\;} (Me_3SiO)_2\overset{\overset{O}{\|}}{P}-O\left[\overset{\overset{O}{\|}}{\underset{\underset{Me_3SiO}{|}}{P}}-O\right]_n\overset{\overset{O}{\|}}{P}(OSiMe_3) \longrightarrow$$

$$\xrightarrow[-Me_3SiOH]{H_2O,\,ROH} RO-\overset{\overset{O}{\|}}{P}(OH)_2$$

(137)

$$R = Alk,\,Ar$$

The reaction of hexaalkyl (aryl) disiloxanes with P_4O_{10} is said to be an efficient synthetic approach to trisilylphosphates which are obtained in high yields (about 80%) [406, 407, 409].

It should be pointed out that the above-mentioned studies were carried out with little spectroscopic evidence. The situation with a good number of reactions of phosphoric anhydride with organosilicon compounds was somewhat similar to that of Langheld-ester when the earlier results had to be revised because nuclear magnetic resonance showed the presence of complicated mixtures of products (Section 1.3). Authors of the publication [409] noted justifiably that up to the late 1960s it was difficult to determine the structures of organosilicon phosphates on the basis of spectroscopic data because IR spectra were not very informative particularly in the cases of polymer formation.

In 1982 Yamamoto and Watanabe [377] provided evidence for the structure of the products of the interaction between phosphoric anhydride and hexamethyldisiloxane. Prolonged heating of the reaction mixture in a sealed NMR tube under an atmosphere of argon was monitored by means of ^{31}P NMR. It was found that no tris(trimethylsilyl) phosphoric acid was formed. This was confirmed by another study [410] and differed from earlier data [405–409, 411]. A small quantity of tetra-(trimethylsilyl) pyrophosphoric acid [408] presence was noted in the mixture.

The reaction was envisaged [377] as proceeding via formation of cyclotetrametaphosphoric acid and hexasilyltetraphosphoric acid (Scheme 138).

Eventually the reaction mixture consisted of the compounds (III–VI) (Scheme 138), the ratio of the products being determined by means of phosphorus-31 NMR. In determining the structures of the products the results were correlated [377] with the data presented in paper [76] rather than those of publication [156]. The different types of phosphorus atoms were assigned to the following signals (δ, ppm):

$$c. -41(P_{branched}), \; -38(P_{middle}) \text{ and } -32(P_{terminal}).$$

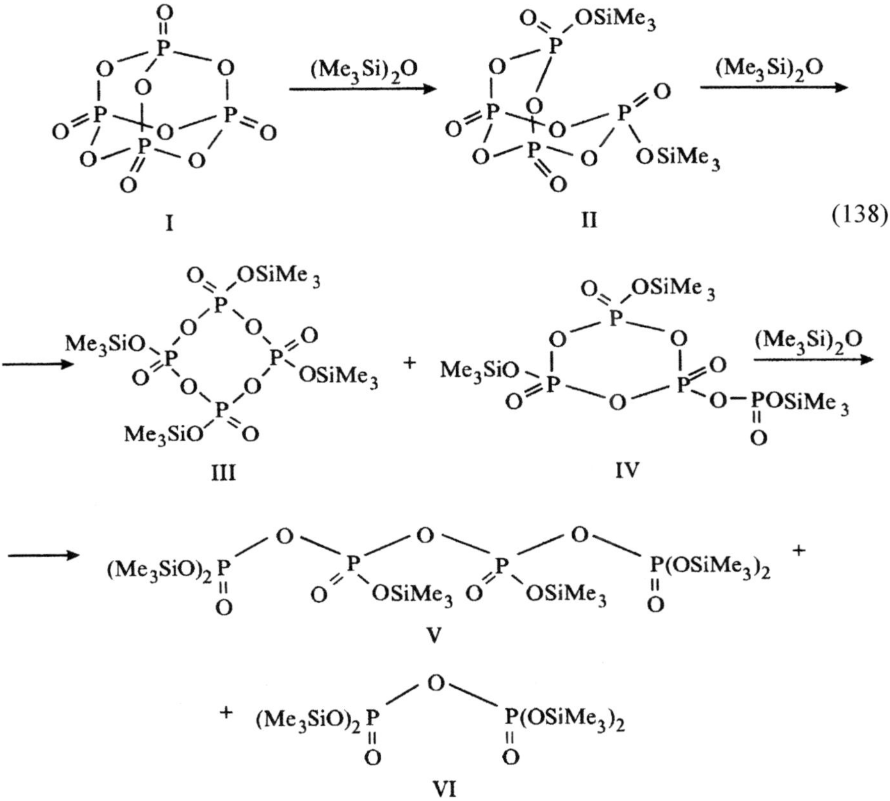

(138)

The characteristic feature of the resulting reaction mixtures, the so-called 'polyphosphoric acid trimethylsilyl ester' (PPSE), was its irregular composition and its strong dependence on solvent and period of heating [377]. A neat sample (without a solvent) contained predominantly compound (V) which gave it a lower reactivity due to the decrease of the branched phosphorus atoms. However, in chlorinated solvents, isotetrametaphosphate (IV) and in xylene tetrametaphosphate (III) structures predominated. These results [377] led to the establishment of the optimum conditions for the synthesis of polyphosphoric acid trimethylsilyl ester with the maximum content of the branched phosphorus atoms that were particularly active in condensation reactions. Such favourable conditions included prolonged heating (24–36 h) at 80°C in a haloalkane (dichloromethane or dichloroethane).

Sulphur and nitrogen-containing analogues of hexamethyldisiloxane reacted with phosphoric anhydride via consecutive breaking up the P–O–P bonds [412] in accordance with Scheme 139.

$$(Me_3Si)_2X \xrightarrow[-(Me_3SiO)_3P=O]{P_4O_{10}}$$

$$X = N \longrightarrow (Me_3SiO)_2 \underset{\underset{O}{\|}}{P}NH(\underset{\underset{O}{\|}}{P}NH)_n\overset{NH-SiMe_3}{\underset{\underset{O}{\|}}{P}}(OSiMe_3)_2 \quad +$$

$$+ \; (Me_3SiO)_2\underset{\underset{O}{\|}}{P}-O(\overset{}{\underset{\underset{OSiMe_3}{|}}{\underset{\underset{O}{\|}}{P}}}-O)_n\underset{\underset{O}{\|}}{P}(OSiMe_3)_2$$

$$X = S \longrightarrow (Me_3SiO)_3P=S \;+\; P_4S_{10}$$

$$(139)$$

It was considered [412] that the reaction proceeded via formation of compounds of the type $(Me_3SiO)_2P(O)XSiMe_3$. This was proved by a model reaction of hexamethyldisylazane and silane with tetraalkylpyrophosphate in accordance with Scheme 140.

$$(Me_3Si)_2X \;+\; (RO)_2\underset{\underset{O}{\|}}{P}-O-\underset{\underset{O}{\|}}{P}(OR)_2 \xrightarrow[-(RO)_2\underset{\underset{O}{\|}}{P}OSiMe_3]{}$$

$$\xrightarrow{\hspace{2cm}} (RO)_2\underset{\underset{O}{\|}}{P}X\,SiMe_3$$

$$(140)$$

$$X = NH, S$$

Hexachlorodisiloxane was found to be inert towards phosphoric anhydride even under rigorous reaction conditions [413].

Trialkyl(aryl)alkoxy silanes reacted with phosphoric anhydride readily [411, 414–420] to give tris (trialkylsilyl) or tris (triarylsilyl) phosphoric acid (Scheme 141).

$$12\,R_3SiOR' + P_4O_{10} \longrightarrow 4\,(R_3SiO)_3P=O \;+\; 6\,R'OR' \qquad (141)$$

The yields of products varied depending upon the length of the substituent's R^1 chain, the compounds with longer alkoxy substituents gave much higher yields [R^1=Me (10%), Et (40–50%) and Bu (60–70%)]. An excess of trimethylalkoxy siloxane in the reaction mixture with phosphoric anhydride did not result in a significant increase in the yield. The optimum molar ratio $Me_3SiOH:P_4O_{10}$ was found to be 4:1 [416, 417], but an excess of phosphoric anhydride could result in a secondary reaction of the latter product leading to dialkyl ether formation, which was not desired.

The interesting publications of Kreshkov and Karateev [411, 417] are devoted to the interaction between phosphoric anhydride and trialkylalkoxy silanes with different monochloroalkyl and dichloroalkyl substituents (Scheme 142).

$$2\,n\,Me_2\underset{\underset{\displaystyle CH_2Cl}{|}}{\overset{\overset{\displaystyle O\,Et}{|}}{Si}} \quad + \quad 0.5\,n\,P_4O_{10} \quad \longrightarrow$$

$$(142)$$

$$\longrightarrow \left[\underset{\displaystyle Me_2(CH_2Cl)\,Si\,O}{|} \overset{\overset{\displaystyle O}{\|}}{-P-}\!\!\underset{\displaystyle |}{}\!\!-O-\overset{\overset{\displaystyle O}{\|}}{P}\!\!\underset{\underset{\displaystyle O\,Si(CH_2Cl)Me_2}{|}}{}\!\!-O- \right]_n \quad + \quad 0.5\,n\,Et_2O$$

Besides the products shown in Scheme 142 polymeric products were also isolated. Their formation was explained by the interaction between tris(trimethylsilyl) phosphoric acid and phosphoric anhydride [411]. Polymeric products were also formed in the reaction of dimethyldibutoxysilane with phosphoric anhydride [415, 416] (Scheme 143).

$$3nMe_2Si(OBu)_2 + 0.5nP_4O_{10} \longrightarrow [Si_3P_2O_8Me_6]_n + 3nBu_2O \quad (143)$$
$$n = 7$$

Van Wazer [3] presented the polymeric products [416] as substituted derivatives of isotetrapolyphosphoric acid (**26**).

The authors [416] named the product dimethylpolysiloxane phosphate because of the alternation of the Si–O–P bonds in the chain. The composition of the product underwent considerable modification during its storage, after 60 days its molecular weight increased from 2250 up to 2970 which corresponded to the value $n \approx 8$.

A similar high molecular compound was formed in the reaction of dimethyldiethoxy silane with phosphoric anhydride. At 200°C the product decomposed to $Si(PO_3)_4 \cdot 2H_2O$ [411, 416]. Borisov, Voronkov and Lukevits [409] assigned the polymeric structure of the type $[Si(PO_3)_4]_n$ to this product. It was concluded [415] that in the reactions of phosphoric anhydride with alkoxysilanes an increase in the number of active functional groups attached to silicon resulted in a higher rate of polymerization.

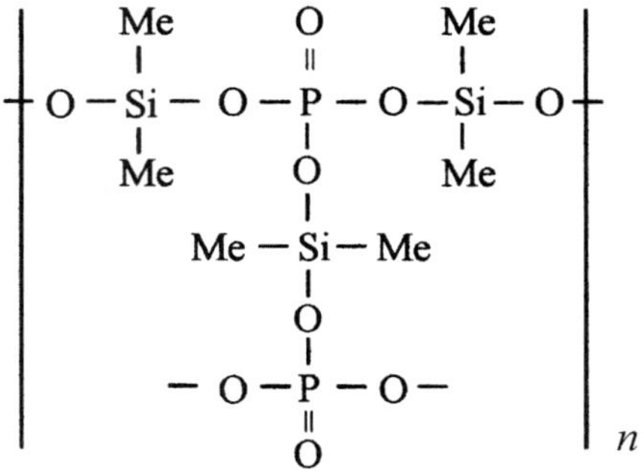

$$
\left[\begin{array}{c}
\overset{\displaystyle Me}{\underset{\displaystyle Me}{O-Si}} - O - \overset{\displaystyle O}{\underset{\displaystyle O}{P}} - O - \overset{\displaystyle Me}{\underset{\displaystyle Me}{Si}} - O \\
Me - Si - Me \\
O \\
- O - P - O - \\
O
\end{array} \right]_n
$$

(26)

Alkyltrialkoxysilanes reacted readily with phosphoric anhydride to give compounds similar to the products formed from dialkyldialkoxy silanes [421, 422] (Schemes 144, 145).

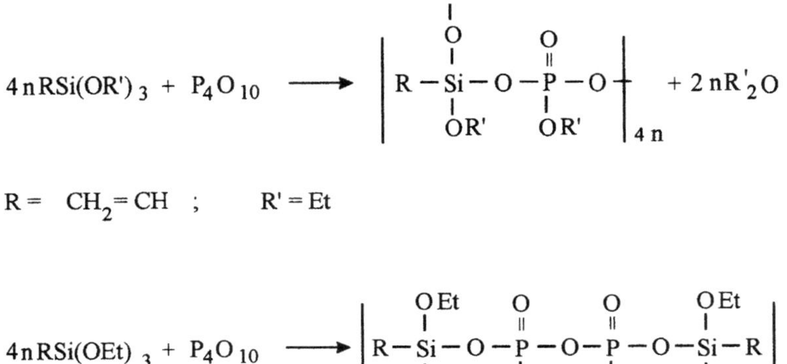

$$
4\,n\,RSi(OR')_3 \;+\; P_4O_{10} \;\longrightarrow\; \left[R - \underset{\displaystyle OR'}{\overset{\displaystyle O}{Si}} - O - \underset{\displaystyle OR'}{\overset{\displaystyle O}{P}} - O \right]_{4n} \;+\; 2\,n\,R'_2O \tag{144}
$$

$$
R = \; CH_2{=}CH \; ; \qquad R' = Et
$$

$$
4\,n\,RSi(OEt)_3 \;+\; P_4O_{10} \;\longrightarrow\; \left[R - \underset{\displaystyle OEt}{\overset{\displaystyle OEt}{Si}} - O - \underset{\displaystyle O}{\overset{\displaystyle O}{P}} - O - \underset{\displaystyle O}{\overset{\displaystyle O}{P}} - O - \underset{\displaystyle OEt}{\overset{\displaystyle OEt}{Si}} - R \right]_{2n} \;+\; 2\,n\,Et_2O
$$

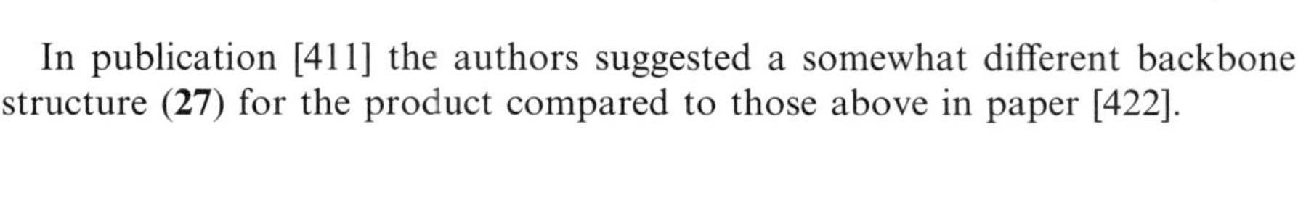

$$
\tag{145}
$$

In publication [411] the authors suggested a somewhat different backbone structure **(27)** for the product compared to those above in paper [422].

$$\left[\begin{array}{cccc} -O & O & O & OR \\ | & \| & \| & | \\ R\,Si-O-P-O-P-O-Si\,R \\ | & | & | & | \\ OR & OEt & OEt & O- \end{array} \right]_n$$

R = Me, Et, Ph

(27)

This structure was strongly criticized by Borisov and co-authors [409]. In the above structure [411] the ethoxy groups are attached to silicon and phosphorus atoms. If the structure was correct the polymer should have been considered as poly-3,5-diphosphono-1,7-tetrasiloxane, not alkylsilyl polyphosphate as was shown by the authors of the publication [421]. The index $2n$ did not convert the formula [422] into the one presented in the paper [411], which corresponded to the structure of the 'head to head' type but not 'head to tail' [409]. The reaction of divinyltetraethoxydisiloxane $[(CH_2{=}CH)(EtO)_2Si]_2O$ with phosphoric anhydride was not accompanied by evolution of an ether [422], which indicated the insertion of phosphorus into the Si–O–Si bonds, and the process of esterification at phosphorus (insertion in the Si–O–C bonds) does not seem reasonable [409]. Authors of the publications [411, 421, 422] did not present any evidence for the alternation of the Si–O and P–O bonds.

Reactions of phosphoric anhydride with tetraaroxy and tetraalkoxy silanes proceeded via transesterification [423]. Tetraaroxy silanes such as tetra-p-tolyloxy silane did not react any further whereas tetraalkoxy silanes gave the products with Si–O–P and P–OAlk bonds. The structures of the polymers produced were determined from the nature of the initial organosilicon compounds [423, 424]. Polymers with similar structures were also produced in the reactions of phosphoric anhydride with tetraethyloxysilane and tetrabutyloxysilane. The number of silicon and phosphorus atoms in the polymeric products had the ratio $1:1$.

A number of relevant examples of the reactions of organic compounds with PPSE are reviewed in Chapter 6, Section 6.4.

The numerous synthetic methods for organosilicon phosphates based on phosphorylating agents other than phosphoric anhydride have been reviewed [409]. The most typical phosphorylating agents for the preparation of these types of products are H_3PO_4 and $POCl_3$ but it is difficult to decide the comparative advantages and disadvantages of the methods applying phosphoric anhydride, orthophosphoric acid or phosphorus oxychloride because of the lack of kinetic studies of such processes.

Polysiloxanes were phosphorylated by P_4O_{10} at 100°C to give mono- and dihydroxyphosphates [424a].

Phosphoric thioanhydride as well as P_4O_{10} reacted with trimethyl(alkoxy) silanes under mild conditions (20–50°C) for 6.5–12 h to produce S-trimethylsilyl-O,O- dialkylthio phosphoric acids along with bis(trimethylsilyl) sulphide [425] (Scheme 146).

$$P_4S_{10} + 8 \ AlkOSiMe_3 \longrightarrow 4 \ (AlkO)_2P(S)SSiMe_3 + 2 \ (Me_3Si)_2S \qquad (146)$$

Still higher yields of the products were achieved with trimethyl(alkenyl) silanes ($c.\,80\%$) and trimethyl(cycloalkenoxy) silanes (85–91%).

The interaction between P_4S_{10} and trimethyl(alkylthio) silanes resulted in the formation of trimethylsilyl esters of dialkyltetrathiophosphoric acids [426] (Scheme 147).

$$P_4S_{10} + 8 \ AlkSSiMe_3 \longrightarrow 4 \ (AlkS)_2PSSiMe_3 + 2 \ (Me_3Si)_2S \qquad (147)$$

Disilylated aliphatic and aromatic diols reacted with thiophosphoric anhydride rather specifically leading to the products of cyclization — the corresponding S-trimethylsilyldithio phosphates [427] (Scheme 148).

$$P_4S_{10} + 4 \ R(OSiMe_3)_2 \longrightarrow 4 \ R \underset{O}{\overset{O}{<}} \!\! \underset{\underset{S}{\|}}{P}\!-\!S\!-\!SiMe_3 + (Me_3Si)_2S \qquad (148)$$

The above reactions of P_4S_{10} with silylated diols and triols illustrate a new and very simple synthetic approach to trimethylsilyl cyclic and acyclic dithio- and tetrathiophosphoric acids. Another attractive point of these reactions from a manufacturing point of view is that no ecologically unfriendly hydrogen sulphide is given off.

7 REACTIONS OF PHOSPHORIC ANHYDRIDE WITH ORGANOMETALLIC COMPOUNDS

Triethylborate reacted with phosphoric anhydride to give an esterified polyphosphate [428] (Scheme 149).

$$n\,(\,BEt_3 \;+\; P_4O_{10}\,) \longrightarrow (EtO)_2\,\overset{\overset{O}{\|}}{P}\left[O-\overset{\overset{O}{\|}}{\underset{\underset{OEt}{|}}{P}}\right]_n O-\overset{\overset{O}{\|}}{P}\,(OEt)_2 \qquad (149)$$

This polymer is produced in industry in significant quantities because it is used as a catalyst in the production of polyacetals.

Another example of esterification of phosphoric anhydride by an organometallic compound is its reaction with aluminium trialcoholate [429] (Scheme 150). The process is carried out at 250°C for 2 h to produce trialkyl or triaryl phosphate.

$$Al(OR)_3 + P_4O_{10} \longrightarrow (RO)_3P{=}O \qquad (150)$$
$$R = Alk,\ Ph$$

Schmidt and Ruidisch [430] studied the reaction of phosphoric anhydride with hexamethyldigermanyloxane. The Ge–O–Ge bond was reactive towards P_4O_{10}, and similar to the corresponding disiloxanes [405–409], tris (trimethyl-germanyl) phosphate was synthesized (Scheme 151).

$$Me_3Ge{-}O{-}GeMe_3 \xrightarrow{\;P_4O_{10}\;} (Me_3GeO)_3\,P{=}O \xleftarrow[-AgCl]{\;Ag_3PO_4\;} Me_3GeCl \qquad (151)$$

$$(Me_3GeO)_3\,P{=}O \xrightarrow{\;H_2O,\ 102°C\;} Me_3GeOH + H_3PO_4 \qquad (152)$$

The structure of trigermanylphosphate was substantiated by its model synthesis (Scheme 151) from germanium trimethylchloride and silver phosphate. The product was rather stable under normal conditions but was hydrolysed by moisture quickly at high temperatures [430] (Scheme 152).

The literature is rich in information on reactions of phosphoric thioanhydride with organometallic compounds compared to phosphoric anhydride reactions. Russian chemists [431–436] have made a large contribution to the target-orientated synthesis of organometallic derivatives of phosphorus thioacids (Scheme 153).

$$P_4S_{10} + 8\,RXER'_3 \longrightarrow 4\,(RX)_2PSER'_3 + 2\,(R'_3E)_2S$$
$$R = Alk;\ R' = Alk,\ Ph;\ X = O,\ S;\ E = Si,\ Ge,\ Sn,\ Pb \qquad (153)$$

The reactions were performed usually in benzene at room temperature, or with heating up to 80°C for 0.5–6 h [433–436]. Under more vigorous conditions (100–140°C for 30–40 min) bis(trialkylstannyl) sulphides reacted with thiophosphoric anhydride leading to the formation of tris (trialkylstannyl) tetrathiophosphates (Scheme 154).

$$P_4S_{10} + 6(R_3Sn)_2S \longrightarrow 4(R_3SnS)_3PS$$

$$R = Me, Bu \tag{154}$$

A good number of the compounds produced in such a way exhibited an interesting range of biological activity.

A general comparison of the chemical properties of phosphoric anhydride and phosphoric thioanhydride was made by Corbridge [1].

8 ELIMINATION, CONDENSATION AND REARRANGEMENT REACTIONS PROMOTED BY PHOSPHORIC ANHYDRIDE

8.1 DEHYDRATION

The importance of phosphoric anhydride and its derivatives such as polyphosphoric acid, tetraphosphoric acid and polyphosphoric esters as exceptionally effective intermolecular and intramolecular dehydrating agents in numerous chemical processes cannot be overestimated. Indeed, phosphoric anhydride and sodium are considered as the two best drying agents [437] for chemical experiments (see Table 3.5). Plesch [437] proposed a very simple and effective method for obtaining pure phosphoric anhydride which he used to study its drying properties.

Table 3.5 Drying ability of the most widely used drying agents [60]

Drying agent	Residual water vapour of gas dried at 298 K	
	(g/litre gas)	(ppm, v/v)
Argon filter (79 K)	1.6×10^{-26}	2.2×10^{-20}
P_4O_{10}	$<2 \times 10^{-8}$	$<2.7 \times 10^{-2}$
$Mg(ClO_4)_2$	$<5 \times 10^{-7}$	<0.68
H_2SO_4 (95%)	9.5×10^{-7}	1.3
Molecular sieve	$c.\,1.5 \times 10^{-6}$	2.0
Silica gel	$c.\,1.5 \times 10^{-6}$	2.0
$Mg(ClO_4)2 \cdot 3H_2O$	$<2.0 \times 10^{-6}$	<2.7
KOH (fused)	2.0×10^{-6}	2.7
BaO	3.0×10^{-6}	4.1
Al_2O_3	3.0×10^{-6}	4.1
$CaSO_4$	4.0×10^{-6}	5.0
MgO	8.0×10^{-6}	10.9
H_2SO_4 (80%)	1.5×10^{-4}	2×10^2
NaOH (fused)	1.6×10^{-4}	2.2×10^2
CaO	2.0×10^{-4}	2.7×10^2
$CaCl_2$ (gran.)	2.0×10^{-4}	2.7×10^2
$CuSO_4$	1.4×10^{-3}	1.9×10^3

Phosphoric anhydride is not only a very powerful dehydrating agent, but it is also widely applicable in organic and bioorganic synthesis. Though a wide range of dehydrating agents is used by chemists in their everyday practice, phosphoric anhydride possesses a number of valuable properties that make it either the best or unique dehydrating agent in many respects. Such properties include very high hydrophilicity in combination with low reactivity towards many functional groups (double and triple bonds, tertiary amines, halogen substituents, carbonyl, nitro, sulphur and some other groups), and mild conditions of the processes. In addition P$_4$O$_{10}$ can produce a very convenient protecting and leaving group which is of particular importance in bioorganic synthesis. Actually the dehydrating activity of phosphoric anhydride is its most well-known and applicable chemical property. It is likely that most of the dehydration and dealkoxylation reactions of phosphoric anhydride, polyphosphoric acid and their derivatives proceed via consecutive steps of phosphorylation–dehydroxyphosphorylation (dealkoxyphosphorylation).

There are many publications on the dehydration of organic compounds promoted by phosphoric anhydride. Thus a limited number of reactions are presented here, predominantly those that are important for organic and bioorganic synthesis, or are less familiar and/or related to the processes described in earlier parts of this book.

In 1890 Kipping [438] presumed that phosphoric anhydride acted as a dehydrating agent in a mixture with lauric or palmetic acid at 200–230°C in a reaction performed without a solvent.

The first (1906) low yield synthetic approach to carbon suboxide involving treating malonic ester with an excess of phosphoric anhydride (dealcoholysis, Sections 1.6 and 8.2) was described by Diels and Wolf [162]. A dehydration method was developed later by Diels and Meyerheim [439, 440]. Carbon suboxide was prepared [439] by the treatment of malonic acid with phosphoric anhydride (Scheme 155).

$$H_2C \begin{matrix} \diagup COOH \\ \diagdown COOH \end{matrix} \quad \xrightarrow[-H_2O]{P_4O_{10}} \quad O=C=C=C=O \qquad (155)$$

The reaction mixture containing malonic acid and a 10–30 fold excess of the dehydrating agent (phosphoric anhydride) was heated to 140–150°C. The low yield of carbon suboxide (12%) was explained [136] by side processes leading to formation of acetic acid, carbon dioxide and polymerization of carbon suboxide. Carbon suboxide [439] was collected as a heavy colourless liquid with boiling point +7°C and a temperature of crystallization between −107 and −111°C. Chemists paid detailed attention to carbon suboxide and studied its reactions with water, amines, hydrazines, halogens, hydrogen sulphide and others in depth [441–444]. Recently Huntress and co-authors [445] presumed that there is carbon suboxide on Halley's comet which provoked even higher interest in this compound. Staudinger made a significant input in working out new synthetic methods for carbon suboxide and ketenes avoiding the use of phosphoric anhydride. The experimental data were presented by him in an impressive set of publications, for example see [160, 446, 447] and the corresponding references in [440].

Half a century later German chemists [412, 448–450] developed a synthetic method for polyfluorosulphur and polyfluoroseleno mono and disubstituted ketenes by the use of phosphoric anhydride for the dehydration of the corresponding substituted acetic acids (Schemes 156, 157).

$$F_5S-\overset{\overset{\displaystyle R}{|}}{C}H-COOH \xrightarrow[-H_2O]{P_4O_{10}} F_5S-\overset{\overset{\displaystyle R}{|}}{C}=C=O \qquad (156)$$

$$R = H,\ Me,\ Cl,\ Br$$

$$(F_3CE)_2CHCOOH \xrightarrow[-H_2O]{P_4O_{10}} (F_3CE)_2C=C=O \qquad (157)$$

$$E = S,\ Se$$

The process was carried out in a column filled with a mixture of sand and phosphoric anhydride. The acid was passed through the column at a very low rate and the temperature of the process maintained between 60 and 160°C. The monosubstituted pentafluorosulphur ketene is a liquid with boiling point 47°C. Its proton NMR spectra contained a doublet with the chemical shift 5.5 ppm (J_{H-F} 5.7 Hz). The authors [448–451] pointed out the high stability of the ketene and its methyl and phenyl analogues at room temperature. It reacted readily with bromine, water, ethanol, hydrogen chloride and alkenes. On short-term heating the ketenes underwent dimerization (Scheme 158).

$$2\,F_5S-CR=C=O \longrightarrow \begin{array}{c} F_5S-CR=C-O \\ |\qquad\quad| \\ F_5S-CR-C=O \end{array} \qquad (158)$$

$$R = H,\ Me,\ Ph$$

The structure of the dimer was determined by mass spectrometry [452]. The presence of the active fluorides (KF, CsF) at high temperature initiated polymerization of ketene. Heating of the pentafluoroketene at 270–290°C resulted in its isomerization to tetrafluorosulfonylideneacetylfluoride (Scheme 159).

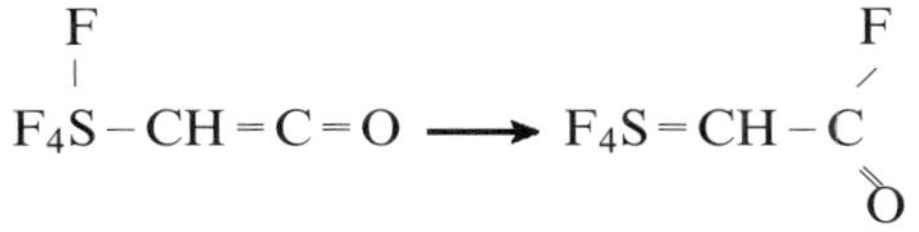

$$F_4S-CH=C=O \longrightarrow F_4S=CH-C\overset{F}{\underset{O}{\diagup}} \qquad (159)$$

Kim and co-authors [453] used phosphoric anhydride efficiently in the synthesis of various acyclic and heterocyclic perfluoro compounds. 2-Oxa- and 2-thia-perfluoroglutaric acids produced in the oxidation of 3,4-dichloroperfluoro-2,5-dihydro furan or the corresponding thiophene, respectively were introduced in the highly exothermic reaction with an excess of phosphoric anhydride to give 2-oxaperfluoroglutaric or 2-thiaperfluoroglutaric anhydride [453] (Scheme 160).

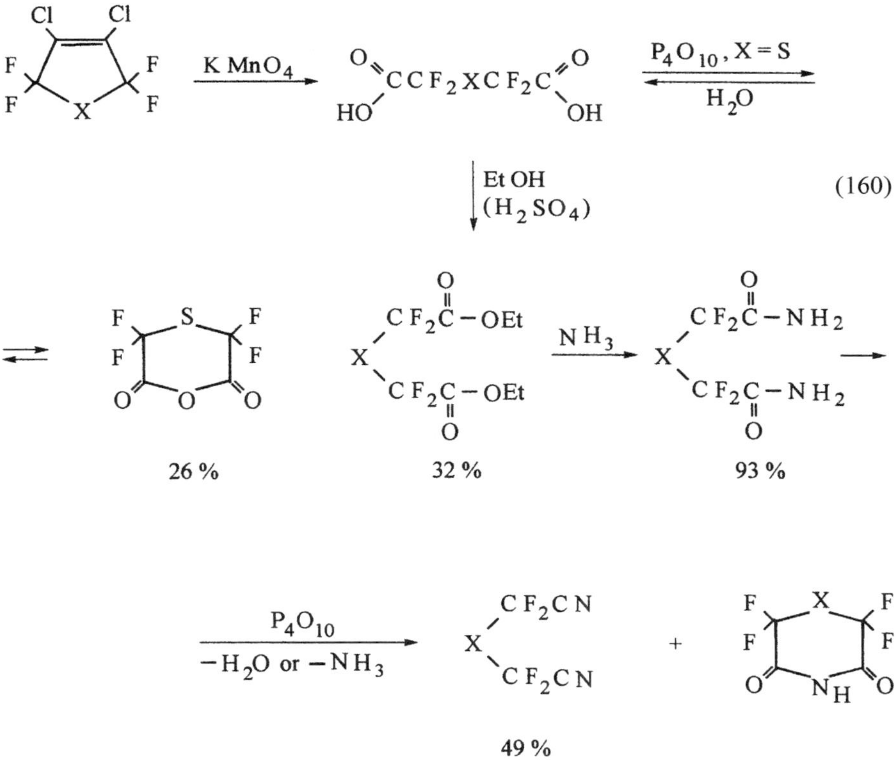

$$ (160) $$

26 %   32 %   93 %

49 %

If the oxidation process of 3,4-dichloroperfluoro-2,5-dihydro furan or thiophene was carried out by one-pot addition of ethanol and concentrated sulphuric acid the corresponding dialkyl esters were produced (Scheme 160). The following treatment of either product by ammonia in ether gave diamide of 2-oxa- or 2-thia-perfluoroglutaric acids. Interaction between a diamide and hot (200°C) phosphoric anhydride under reduced pressure or in the atmosphre of nitrogen resulted in the formation of the corresponding 2-oxa- or 2-thia-perfluoroglutaro nitrile [453] (Scheme 160). The last step in the synthesis was accompanied by partial deamination-cyclization in accordance with Scheme 160.

An efficient intramolecular dehydration of N,N′-disubstituted urea with phosphoric anhydride was studied by Stevens and co-authors [454, 454a]. A four-fold excess of P_4O_{10} reacted with urea in pyridine in the presence of sand under reflux gave the corresponding carbodiimides (Scheme 161) with the yield of 75%.

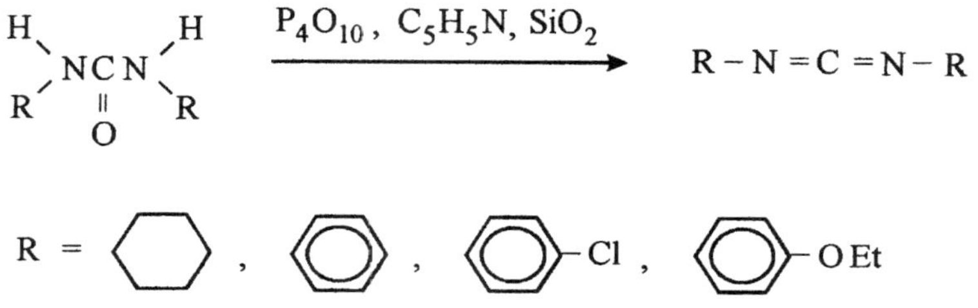

$$R - N = C = N - R \qquad (161)$$

Probably pyridine acted in this process as a nucleophilic promoter of the interaction (Section 3.2). It is interesting to compare the above data with those published by Van Wazer and co-authors [8] and reviewed earlier (Chapter 1).

A useful synthetic approach to keteneimines involves the dehydration of the corresponding amides. Unlike acetanilides and benzanilides (Sections 3.3.2 and 3.3.3) diphenylacetanilides suffered dehydration under the action of phosphoric anhydride. The reaction proceeded under specific conditions: it was carried out in pyridine with 5–10-fold molar excess of phosphoric anhydride and 20–30-fold molar excess of aluminium oxide [226, 454–456] (Scheme 162).

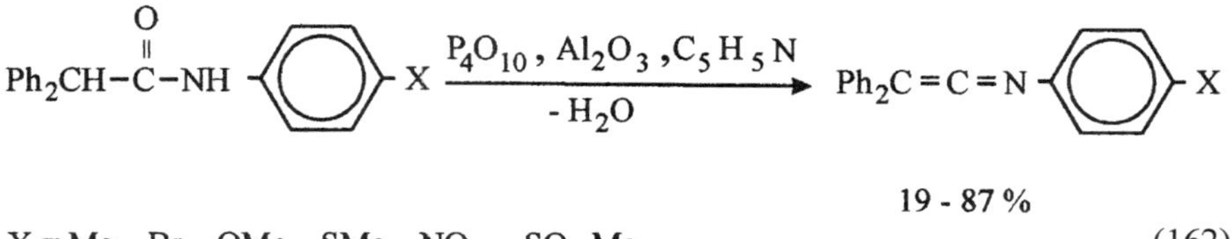

19 - 87 %

$$X = Me, \ Br, \ OMe, \ SMe, \ NO_2, \ SO_2Me \qquad (162)$$

Phosphoric anhydride has been used successfully for the syntheses of a wide range of nitriles from amides. Numerous fluorine-containing nitriles were prepared in such dehydration reactions. For example, perfluoroalkyl amides, acyl perfluoroayl amides and perfluoro vinylesters amides reacted with phosphoric anhydride at high temperature (160–200°C) to give corresponding perfluoroalkyl nitriles [457], acyl perfluoroacyl amides [458] or perfluoro vinyl ether nitriles and their derivatives [459] (Scheme 163).

$$R(CF_2)_n\underset{\overset{\|}{O}}{C}NH_2 \xrightarrow[-H_2O]{P_4O_{10}} R(CF_2)_nC \equiv N \tag{163}$$

$$R = CF_3, OAc, OCF = CF_2 \quad n = 2, 3, 4, 6$$

From an industrial point of view an attractive method for the preparation of dicyanoacetylene [460] is the dehydration of the diamide of acetylenedicarbonic acid by phosphoric anhydride [460–463] in accordance with Scheme 164. Dicyanoacetylene is a very powerful dienophile and has been used as a source of energy in rockets and lasers.

$$\underset{H_2N}{\overset{O}{\diagdown}}C - C \equiv C - \underset{NH_2}{\overset{O}{C}} \xrightarrow[-H_2O]{P_4O_{10}} N \equiv C - C \equiv C - C \equiv N \tag{164}$$

The technological aspect was studied in depth [461, 462] as a preliminary step for the industrial production of dicyanoacetylene. An alternative dehydration agent, $(COCl)_2$–DMSO which converted amides to monocyano compounds in high yields was described recently [464].

An early example of phosphoric anhydride participation in the amide to nitrile transformations was its solid state heterogeneous reaction with isobutyramide under rigorous conditions that resulted in isobutyronitrile [465, 466] (Scheme 165). Under mild conditions amides are phosphorylated by P_4O_{10} (Section 3.3.1).

$$\underset{Me}{\overset{Me}{\diagdown}}CHC\underset{\diagdown NH_2}{\overset{\diagup O}{}} \xrightarrow[-H_2O]{P_4O_{10}} \underset{Me}{\overset{Me}{\diagdown}}CHC \equiv N \tag{165}$$

$$69 - 86\,\%$$

A similar method was used by Teague and Short [467] for the synthesis of nicotinonitrile in good yield from nicotinamide (Scheme 166).

$$\text{(nicotinamide)} \xrightarrow[-H_2O]{P_4O_{10}} \text{(3-cyanopyridine)} \qquad (166)$$

$$83 - 84\ \%$$

The synthetic approach to ketene acetals was a trial issue for chemists in the 1920–40s. A number of methods were proposed: however, all of these were criticized in one way or another until McElvain and Clarke [468] proposed a multi-step synthesis of ketals from alkyl ester of carboxylic acid. An amide of the corresponding acid was subjected to dehydration into nitrile under the action of phosphoric anhydride and triethyl amine (Scheme 167).

The following consecutive steps via the formation of iminoester hydrochloride and orthoester led to tetraethoxy ethylene in accordance with Scheme 167 [468].

$$\underset{EtO}{\overset{EtO}{>}}CHC\overset{O}{\underset{OEt}{<}} \xrightarrow{NH_3} \underset{EtO}{\overset{EtO}{>}}CHC\overset{O}{\underset{NH_2}{<}} \xrightarrow{P_4O_{10}}$$

$$84\ \%$$

$$\longrightarrow \underset{EtO}{\overset{EtO}{>}}CHC\equiv N \xrightarrow{EtOH,\ HCl} \underset{EtO}{\overset{EtO}{>}}CHC\underset{OEt}{\overset{=NH\cdot HCl}{|}} \longrightarrow$$

$$76 - 79\ \% \qquad\qquad 70\ \%$$

$$\xrightarrow[\text{r.t., 4 days}]{EtOH} \underset{EtO}{\overset{EtO}{>}}CHC(OEt)_3 \xrightarrow{Et\,Na} \underset{EtO}{\overset{EtO}{>}}C=C\overset{OEt}{\underset{OEt}{<}}$$

$$42\ \% \qquad\qquad 39\ \%$$

$$(167)$$

It is noteworthy that polyphosphoric acid is a relatively poor dehydrating agent for converting amides into nitriles, but can be usefully applied to the transformation of nitriles into amides. In the latter reactions phosphoric anhydride is not suitable.

Hydration of nitriles under the action of polyphosphoric acid was studied by Snyder and Elston [469]. Relatively facile reactions took place at 110°C over approximately 1 h, to give good yields of the corresponding amides. The most typical example of the process studied was hydration of benzonitrile. α-Hydroxybutyronitrile displayed some interesting features in its reactivity towards water in the presence of polyphosphoric acid. At room temperature the process occurred smoothly with the formation of α-hydroxybutyramide in a moderate yield [469] (Scheme 167a). Attempts to increase the yield of the reaction at an elevated temperature (85°C) resulted in an elimination–condensation reaction with the formation of 2,2,5,5-tetramethyl-4-oxazolidine [469] (Scheme 167a). This structure was elegantly confirmed by its synthesis from α-hydroxybutyramide and acetone (Scheme 167a).

$$\text{(167a)}$$

Hydration–decarboxylation reaction took place at 160°C in the case of 2,4,6-trimethyl cyanobenzene amide formation being followed by further hydrolysis and decarboxylation with eventual loss of the functional group as expressed by Scheme 167b [469].

However, this synthetic method is inapplicable to highly sterically hindered nitriles. For example under similar conditions 2,4,6-triisopropyl benzonitrile and 1-hydroxy-2-cyano-3-methyl naphthalene did not undergo the corresponding transformations [469] (Scheme 167c).

The remarkable potential of polyphosphoric acid in synthesis was exhibited by Hauser and Murray [470]. The starting point for the synthesis of 2-methyl-3-phenyl-4-quinolinol was the conversion of β-ketonitrile into β-ketoamide under the action of unspecified polyphosphoric acid at 100°C without a solvent [470] (Scheme 168). The isolated α-phenylacetoacetamide was condensed with aniline in the presence of acetic acid to produce the corresponding anil-amide. The latter compound was cyclized into quinolinol again under the action of polyphosphoric acid at high temperature (140°C) [470] (Scheme 168).

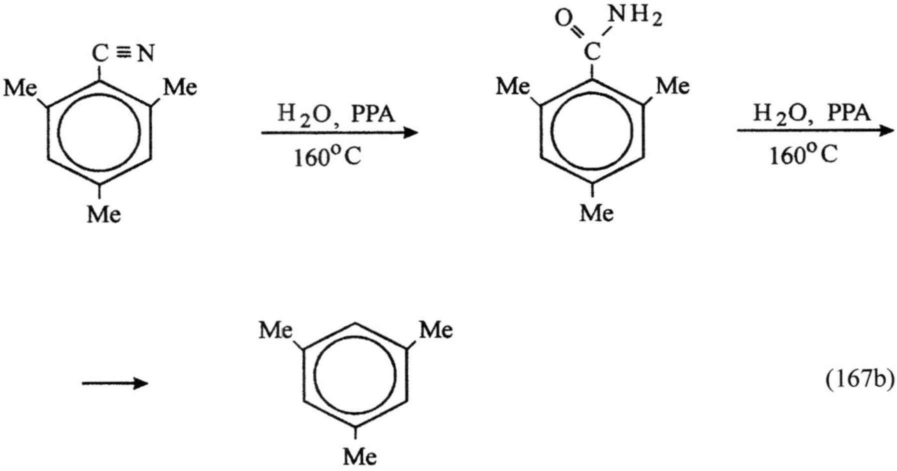

(167b)

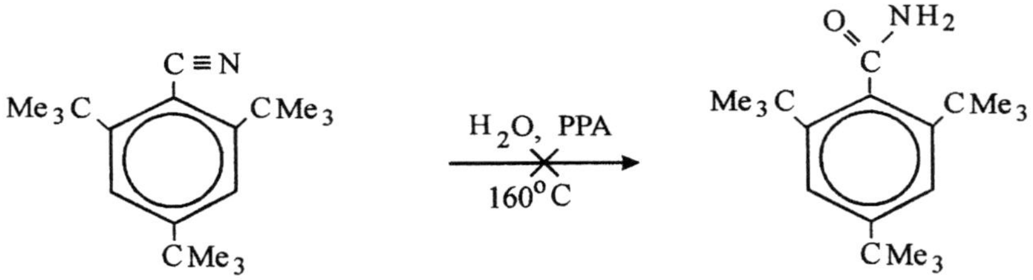

(167c)

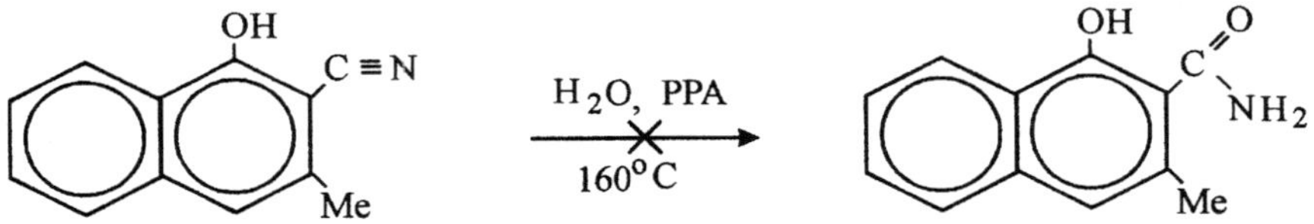

A combination of phosphoric anhydride and tertiary amine was found to have a pronounced dehydrating activity and thus was particularly useful for the dehydation of acid-sensitive amides. Dehydration of amides by the action of phosphoric anhydride possesses clear advantages over phosphorus pentachloride, since the latter reagent also produces chlorinated side products.

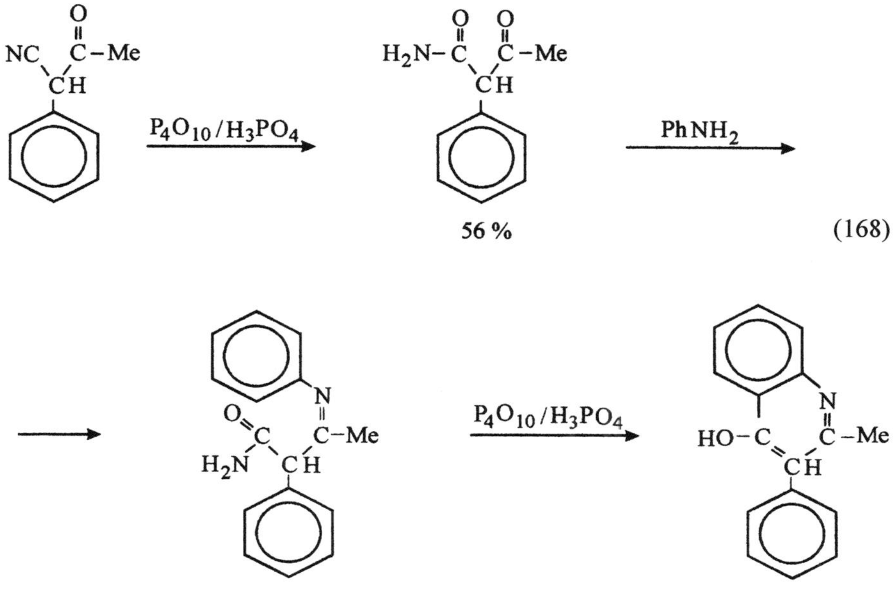

$$(168)$$

Thiophosphoryl anhydride exhibited no potential for the dehydration of amides into the corresponding nitriles but instead substituted the amido oxygen with sulphur. The resulting thioacetamide was condensed with chloroacetone under refluxing in benzene to produce 2,5-dimethylthiazole [471, 472] (Scheme 169).

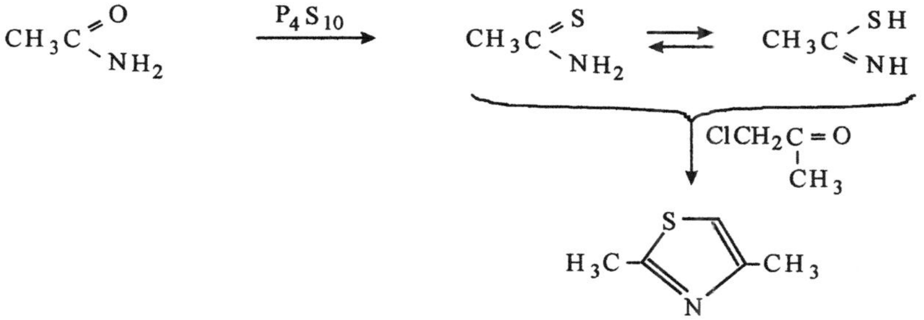

$$(169)$$

A very useful version of amide dehydration–condensation is the Bischler–Napieralski reaction known since 1893 [473]. The process of self-condensation in N-acyl-β-arylethylamines took place under the action of phosphoric anhydride in boiling xylene or preferably tetralin and resulted in the formation of the dihydroisoquinoline shown below [226, 474] (Scheme 170).

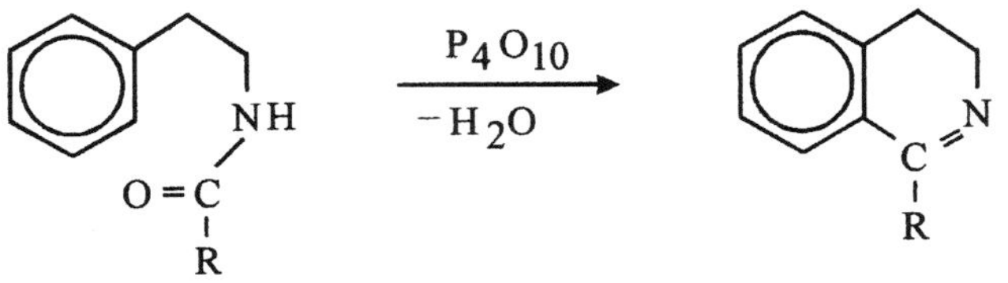

$$(170)$$

Phosphoric anhydride was also used as a cyclisation reagent for the synthesis of a substituted indole at high temperature with moderate yield in accordance with Scheme 170a [475, 476].

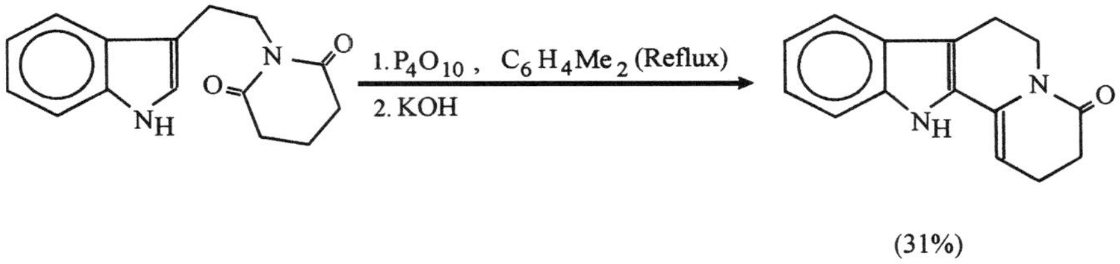

(31%)

$$(170a)$$

Itoh and Sugasawa [477] worked out a method for dehydration–cyclization of β-(3,4-dimethoxyphenethyl) acetamide by treating it with a mixture of phosphoric anhydride with sand in boiling pyridine. The product, 6,7-dimethoxy-1-methyl-3,4-dihydroisoquinoline, was isolated in high yield (85%).

The Bischler–Napieralski reaction was initiated also by the derivatives of phosphoric anhydride such as polyphosphoric ester (PPE) and polyphosphoric acid (PPA). The former reagent was prepared from P_4O_{10} and diethyl ether in chloroform (Section 1.3) and introduced in the reaction with amides *in situ* [478] (Scheme 170b) to produce the products similarly to P_4O_{10} (Scheme 170).

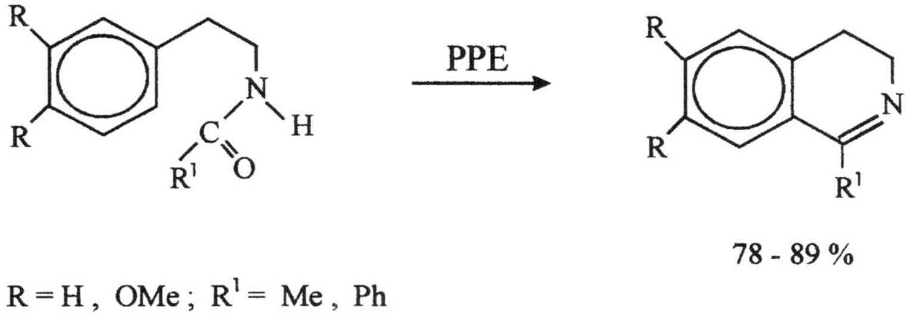

$$(170b)$$

78 - 89 %

$R = H$, OMe ; $R^1 = Me$, Ph

The process was performed under mild conditions in the presence of 5-fold excess of polyphosphoric ester. The methoxy substituents in the aromatic ring enhanced the reaction rate which resulted in 78–85% yield of the products of cyclization (dihydroisoquinolines) after 30 min of refluxing in the chloroform solution. The other phenylethylamides were somewhat less reactive but nevertheless their 60–90 min heating with polyphosphoric ester at 120°C led to the corresponding products which were isolated in high yields [478].

γ-(3,4-Dimethoxyphenyl) propylamides under the action of polyphosphoric ester were cyclized into 3,4-dihydrobenzazepines also at 120°C and consequently reduced by sodium borohydride with the formation of 1,2,3,4-tetrahydroazepines [478] (Scheme 171).

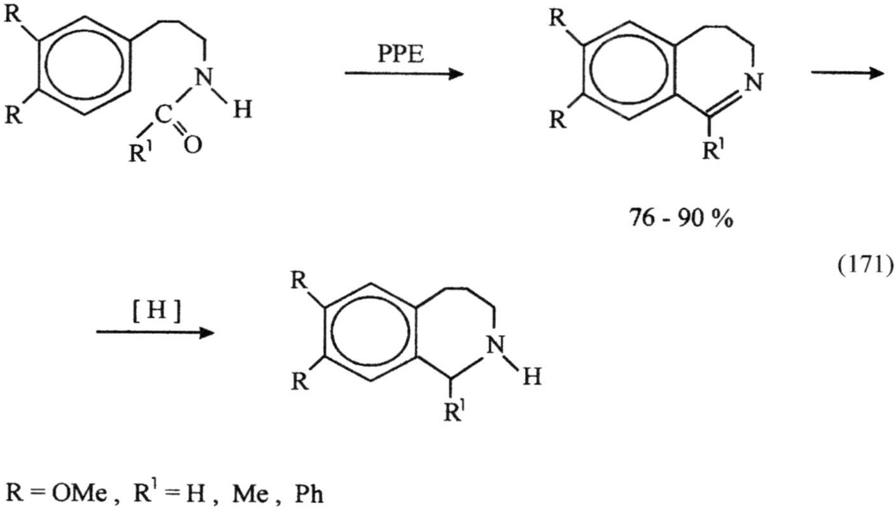

76 - 90 %

(171)

R = OMe , R¹ = H , Me , Ph

Bailey and co-authors [478a] delivered a one-pot synthesis of tetrahydroquinolines from alkylaryl amine and formic acid involving polyphosphoric acid enriched with phosphoric anhydride as an initiator. The Bischler–Napieralski condensation reaction took place during the second step of the process resulting in a very good yield of the product (Scheme 172).

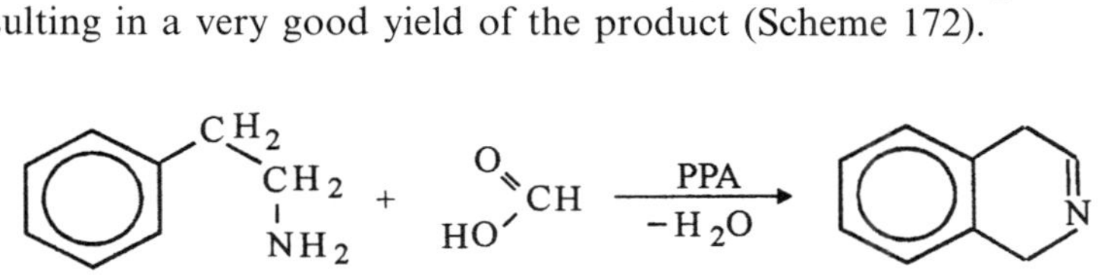

(172)

A process closely related to the Bischler–Napieralski condensation is the Pictet–Gams reaction [479–482] which involves cyclodehydration of β-hydroxy-β-phenylethylamides under the action of phosphoric anhydride, polyphosphoric acid or polyphosphoric ester (Scheme 173).

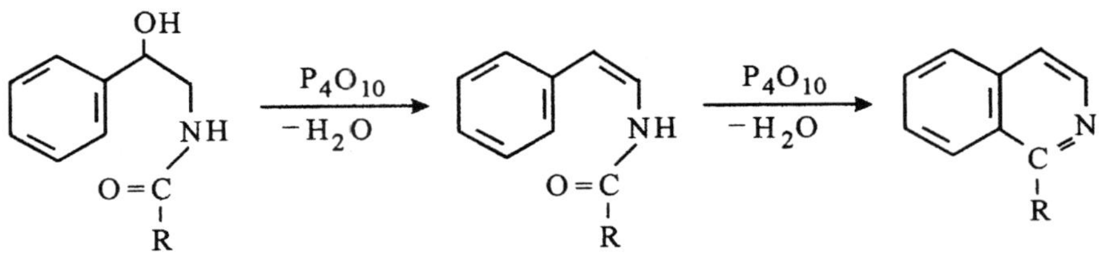

$$(173)$$

A substituent in the α-position to the amide group can migrate to the C4 atom. A typical example of this reaction is the synthesis of Papaverine (see Chapter 5).

The classical mechanism for the above reactions of cyclization which involves formation of the carbocation is expressed by Scheme 174 [483].

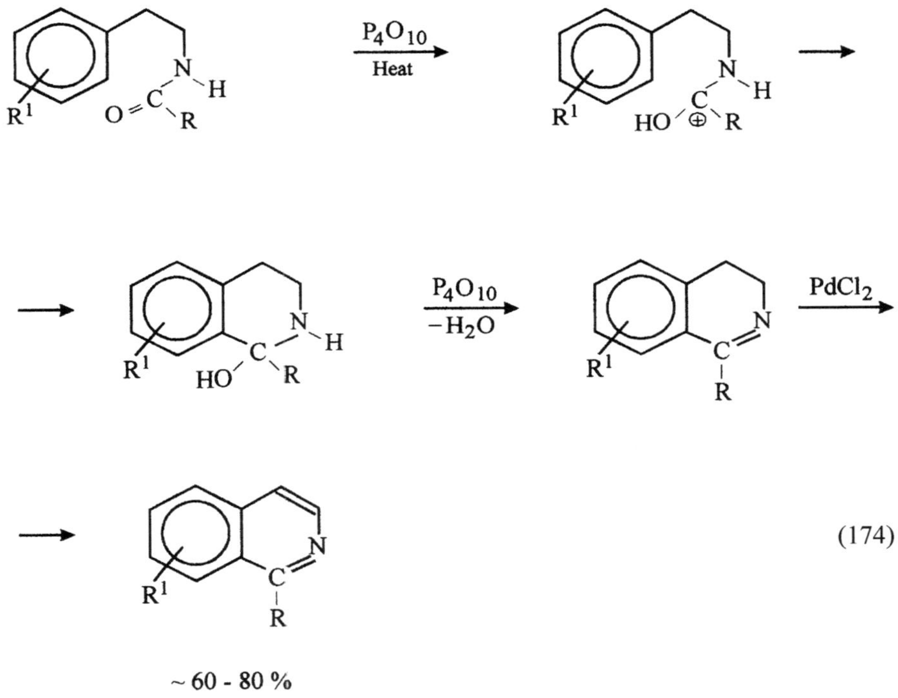

$$(174)$$

$\sim 60 - 80\ \%$

According to Fodor, Gal and Phillips [484] the Bischler–Napieralski ring-closure process involving P_4O_{10} proceeds via an imidoyl phosphate which equilibrates with nitrilium phosphate. Such a mechanism was supported by the synthesis of 6,7-dimethoxy-1-phenyl-3,4-dihydroisoquinoline hydrochloride from N-[2-(3,4-dimethoxyphenyl)ethyl] benzamide as shown in Scheme 175 [484].

$$(175)$$

The reaction involving PCl_5 was monitored by proton NMR [484] which exhibited a distinct non-equivalence of the protons of the two methoxy groups in the product of cyclization (two singlets at 3.80 and 4.09 ppm) whereas in the spectrum of the initial reagent the high-field area contained only one singlet at 3.82 ppm. A new signal at 4.68 ppm instead of the signal at 4.02 ppm was attributed to the methylene group adjacent to nitrogen in the nitrilium ion formed on the reaction pathway [484].

Nevertheless it is possible that the Bischler–Napieralski and Pictet–Gams reactions which are promoted by phosphoric anhydride, polyphosphoric acid or polyphosphoric ester proceed via phosphorylation of the highly nucleophilic nitrogen in amides followed by the process of N⟶O phosphorotropy with the formation of the phosphate which transforms into the dihydroisoquinoline structure upon elimination of the phosphate (Scheme 176) (similar to the pathway of benzoylamides, acetanilides and benzanilides phosphorylation by phosphoric anhydride presented in Sections 3.3.1–3.3.3).

The interest in the Bischler–Napieralski reaction and its versions is pronounced and new initiating agents have been recently proposed for the process [485]. Still phosphoric anhydride and its derivatives are the only efficient initiators in a good number of the Bischler–Napieralski dehydration–cyclization reactions where classical reagents fail [486] and are probably the preferable reagents in this reaction from technological and ecological points of view.

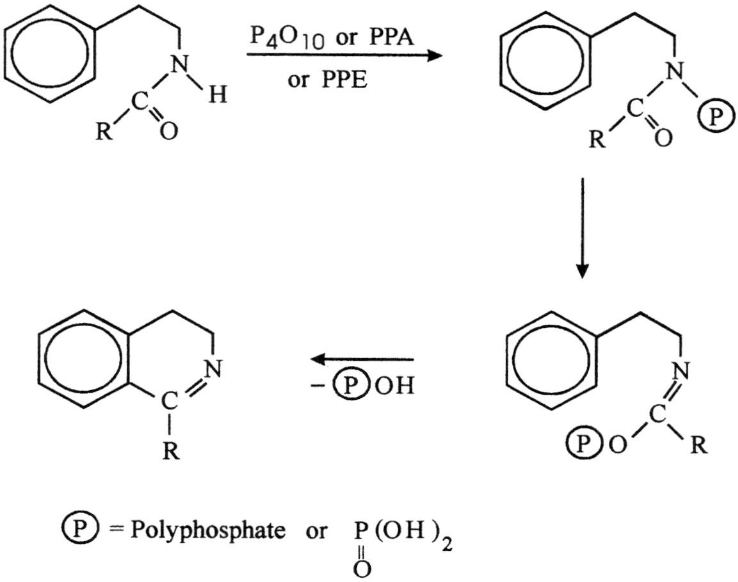

$$\text{(176)}$$

$$\circled{P} = \text{Polyphosphate} \quad \text{or} \quad P(OH)_2 \ (=O)$$

Phosphoric anhydride promotes transacylation in the reaction between dimethylformamide and alkyl or aryl carboxylic acids at the elevated temperature [487] (Scheme 177).

$$R{-}C\underset{OH}{\overset{O}{<}} + Me_2NCHO \xrightarrow[\text{Refluxing}]{P_4O_{10}} R{-}C\underset{NMe_2}{\overset{O}{<}} \qquad \text{(177)}$$

$$R = \text{Alk, Ar}$$

The significant advantage of such a method consists in avoiding the preliminary synthesis of corresponding acid chlorides or anhydrides [488].

Under the action of phosphoric anhydride various arylcarboxylic acids underwent intramolecular dehydration–condensation [489]. In 1932 Cook [490] published the results of such reactions using 2-carboxy-1,1'-dinaphthylketone and some of its analogues. The reaction of this compound with phosphoric anhydride in nitrobenzene at 150°C led to the formation of 1,2,5,6-dibenzantraquinone in high yield (Scheme 178).

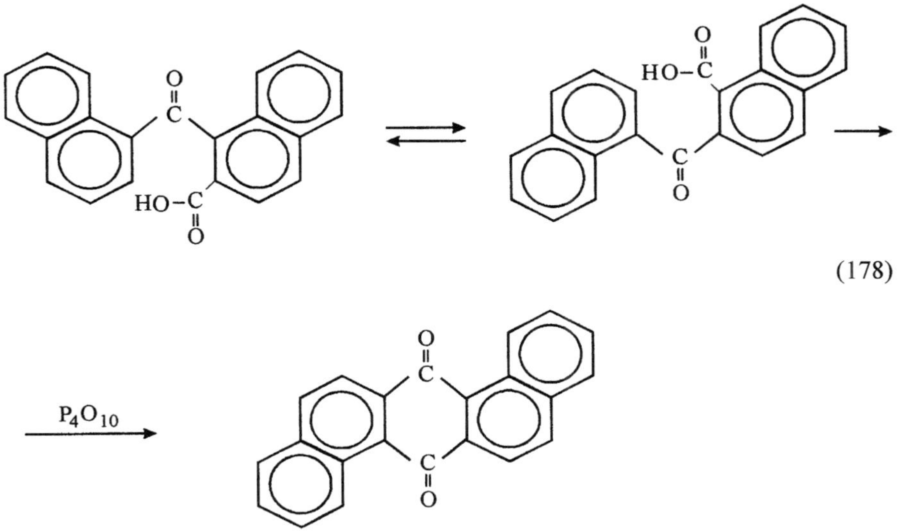

(178)

$$\xrightarrow{\quad P_4O_{10} \quad}$$

The result of the interaction was explained in terms of α-δ migration of the 1-naphthyl group. 2-Carboxy-1,2′-dinaphthyl ketone in the presence of phosphoric anhydride underwent the process of dehydration–cyclization with the partial migration of the naphthyl group to give the mixture of dibenzanthraquinones [490].

A similar effect was observed in the case of the product of the Friedel–Crafts condensation between 1,2-naphthoic anhydride and thiophene. The mixture of isomers was subjected to dehydration–cyclization in the presence of phosphoric anhydride to give the quinone derivative [490a] (Scheme 179).

This type of rearrangement was described sometime later by Bergmann and Szmuszkovicz [491] for 2-methyl-2′-naphthyl-1-tetralone under the influence of phosphoric anhydride. When heated at 135–140°C for 60 min with P_4O_{10} in an atmosphere of nitrogen, tetralone gave not the expected product of dehydration but the compound characterized by the authors [491] as 7,8-dibenzo-1,2,5,6-fluorene (Scheme 180) though without reliable confirmation of its structure.

Dehydration of the amide of α-sulfofluoro tetrafluoropropionic acid has to be singled out because it did not produce the corresponding nitrile, but proceeded further towards a product which was the result of its cyclotrimerization with the high yield (60%). The nitrile was isolated only as a minor product (9%) [457] (Scheme 181).

$$\text{(179)}$$

This is an interesting rare example of π-system cyclotrimerization under the action of P_4O_{10}. Cyclotrimerization of monosubstituted ketenes into the cyclic systems induced by phosphoric anhydride is presented in Section 8.2.

Middleton [492] performed an impressive synthesis of 1,1-dicyano-2,2-bis (trifluoromethyl) ethylene by condensation of perfluoroacetylene with dicyanomethane in the presence of zinc chloride followed by dehydration of the unstable product by phosphoric anhydride (Scheme 182). A similar procedure was applied to the cyclic analogue [492] (Scheme 183).

Phosphoric anhydride is a convenient reagent for the preparation of anhydrides from their acids. Thus the following acids have been converted to their anhydrides — tribromo acetic acid [493], perfluorooctane sulphonic acid [494], 1-bromo- and 1-chloro-tetrafluoroethane sulphonic acid [495]. The anhydride of nitrodifluoro acetic acid has also been prepared using P_4O_{10} [496, 497].

$$(180)$$

$$FSO_2\underset{CF_3}{CF}-\underset{O}{\overset{O}{C}}NH_2 \quad \xrightarrow[-H_2O]{P_4O_{10}} \quad + \quad (181)$$

$$+ \quad FSO_2\underset{CF_3}{CF}-C\equiv N$$

$$CF_3-\overset{O}{\underset{\parallel}{C}}-CF_3 \quad + \quad CH_2(CN)_2 \quad \xrightarrow{ZnCl_2} \quad HOC(CF_2)_2CH(CN)_2 \longrightarrow \quad (182)$$

$$\xrightarrow[-H_2O]{P_4O_{10}\,;\,\text{Distill.}} \quad (CF_3)_2C=C(CN)_2$$

$$50\%$$

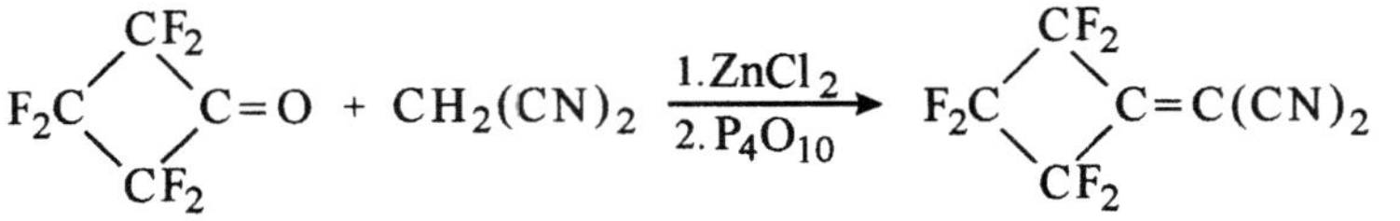

(183)

A good yield of p-toluenesulphonic anhydride was achieved [498] by the action of phosphoric anhydride on p-toluenesulphonic acid for 9 h at 125°C (Scheme 184).

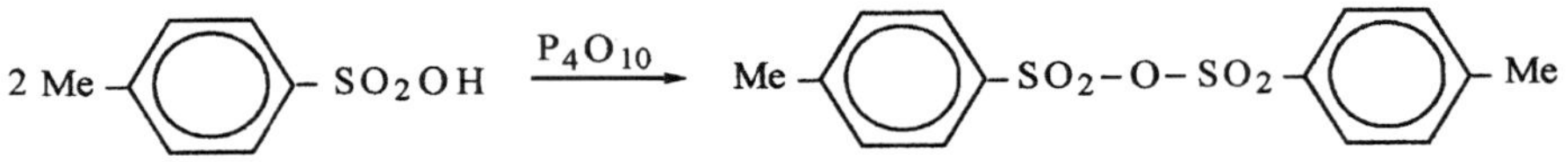

(184)

Phosphoric anhydride was used efficiently on the first step of synthesis of alkenyl trifluorosulfonates from trifluoromethylsulphonic acid [499] in accordance with Scheme 185.

$$CF_3SO_3H \xrightarrow[-H_2O]{P_4O_{10}} (CF_3SO_2)_2O \xrightarrow[\text{Pentane, Pyridine}]{Me_2CHCOMe} Me_2C{=}\underset{\substack{|\\ Me}}{C}{-}OSO_2CF_3 \quad (185)$$

98%

A mixture of phosphoric anhydride with phosphoric acid initiated the transformation of γ-butyrolactone into cyclopentenone [500] (Scheme 186). It is noteworthy that there can be a considerable difference from the point of chemical action between ready-made polyphosphoric acid and a mixture P_4O_{10}/H_3PO_4 the components of which are added in the reactor separately.

An attractive recent example of intermolecular dehydration induced by phosphoric anhydride is a high yield (95%) formation of geranyl formate from geraniol and formic acid at room temperature [501].

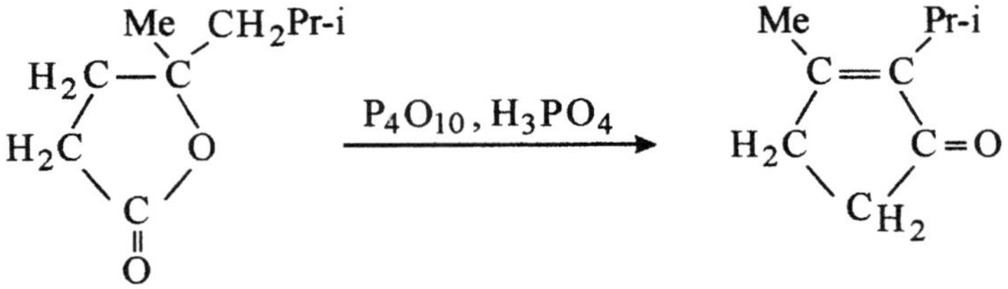

(186)

Recently the Knoevenagel condensation was reported to be initiated by microwave irradiation in the presence of phosphoric anhydride [502]. This method was worked out for p-quinolone methide derivatives, which can be considered as stereometric models of quinolone antibacterials. The condensation took place between diethyl malonic ester and cyclic aldehydes such as benzaldehyde, 1-naphthaldehyde and indol-3-carboxaldehyde in the presence of piperidine (Schemes 187–189). Dealcoholysis of malonic ester (Section 3.8.2) was not observed in these cases.

R = H, Me, OMe ; R' = Cycle

(187)

(188)

(189)

MWI - Microwave irradiation

The microwave irradiation of the reaction mixtures was needed for only 5–15 min to give reasonable yields of the condensation products (47–87%). A longer period of irradiation did not increase the yield. The influence of the ratio cyclic aldehyde:diethyl malonic ester:piperidine was studied [502] but surprisingly the role of phosphoric anhydride was not reduced to the amount only needed to remove the water produced in the reaction. Otherwise it could appear to be much more significant.

The high yield one-pot synthesis of sulphamides was carried out in xylene at 150°C for 3 h [503] (Scheme 190).

$$\text{(190)}$$

The siginificant point of the above synthesis is that phosphoric anhydride acted as an efficient reagent for the condensation reaction prior to dehydration.

The same products could be obtained without phosphoric anhydride but only after 10 h of heating [503]. The participation of phosphoric anhydride in the conceivable transformations in the N-substituted derivatives when the process of dehydration became impossible was also investigated [503] (Scheme 191).

$$\text{(191)}$$

The mixture of phosphoric anhydride with methanesulphonic acid (P_4O_{10}/ MeSO$_3$H) is another efficient dehydrating agent containing phosphoric anhydride which has been introduced and studied within the past three decades. The usual ratio of components is 1:10 by weight, respectively. The mixture of phosphoric anhydride with methanesulphonic acid was found [504] in many respects to be a convenient alternative to polyphosphoric acid as a dehydrating agent. The advantages included easiness to pour and stir due to the low viscosity, high solubility of many organic substrates and moreover it is easy to work-up the products from the reactions carried out in P_4O_{10}/MeSO$_3$H because it decomposes readily when treated with either water or sodium bicarbonate solution.

Though the dehydrating agent $P_4O_{10}/MeSO_3H$ is called 'the mixture' of one component with another, it is probably the product of the chemical interaction between the two compounds. The structure of this product was assigned by Ueda [505] to that shown in Scheme 192.

$$(192)$$

Taking in consideration the ratio of the reagents $P_4O_{10}:MeSO_3H$ ($1:10$) it is possible that the process of the P–O–P bonds cleavage in phosphoric anhydride is proceeding further than the opening of one cycle. On the other hand the acid may simply be protonating P_4O_{10} and acting as a solvent.

The activity of chemical dehydration by the phosphoric anhydride–methanesulphonic acid mixture was illustrated very well [505] by the reactions that had been classically performed with polyphosphoric acid. Olefinic acids or their lactones underwent intramolecular acylation under the action of $P_4O_{10}/MeSO_3H$ with the formation of substituted cyclopentenones (Scheme 193).

$$(193)$$

The Beckman rearrangement of benzophenone oxime also occurred in $P_4O_{10}/MeSO_3H$ upon heating at 100°C for 1 h to give benzanilide [505] (Scheme 194).

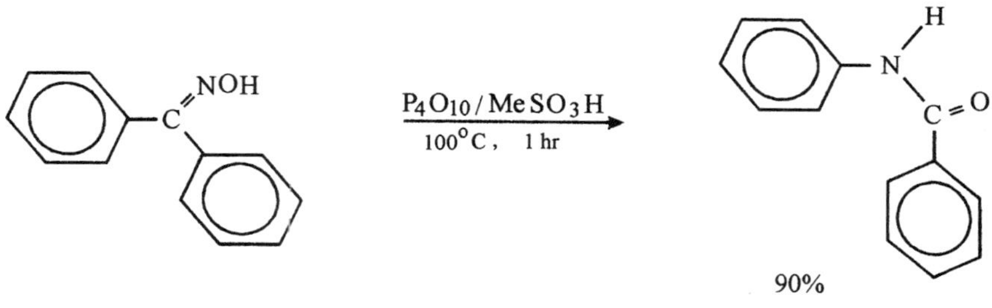

$$\text{(194)}$$

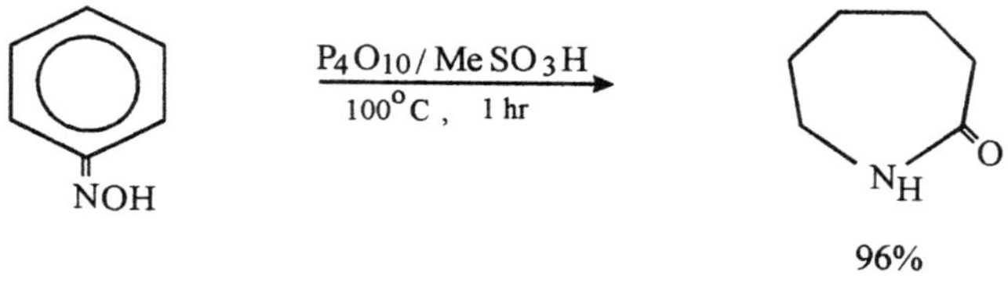

$$\text{(195)}$$

The same mixture $P_4O_{10}/MeSO_3H$ provided an excellent driving force for the Beckman rearrangement of cyclohexanone into ε-caprolactam [505] (Scheme 195).

The yields of dehydration and dehydration–condensation reactions involving polyphosphoric acid and the phosphoric anhydride–methanesulphonic acid mixture were very close, which gave the priority to the latter over polyphosphoric acid only from the point of ease of procedure.

Instead of methanesulphonic acid mixed with phosphoric anhydride trifluoromethylsulphonic acid can be used efficiently. Intramolecular dehydration shown in Scheme 196 took place under the action of P_4O_{10}/F_3CSO_3H used in the ratio $1:20$ by weight [506].

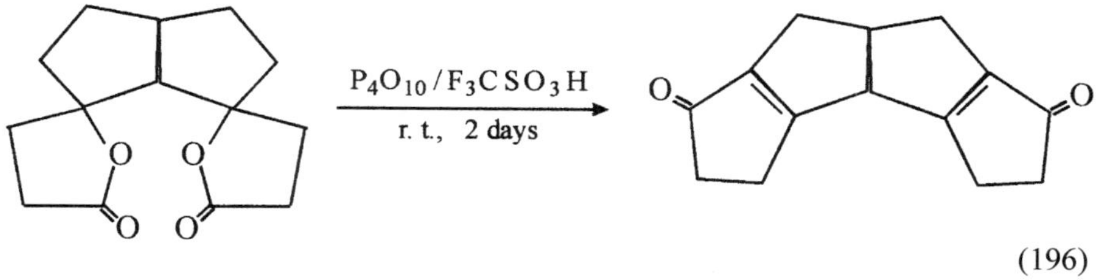

$$\text{(196)}$$

The phosphoric anhydride–trifluoromethanesulphonic acid mixture can be handled without difficulties but its advantages as a dehydrating agent over $P_4O_{10}/MeSO_3H$ in this case as well as in some other processes are not obvious.

A significant input into the scope of the reactions carried out in $P_4O_{10}/$ $MeSO_3H$ was made by Ueda and co-workers. A wide range of substituted 2-benzimidazoles and benzothiazoles were synthesized by the reaction of o-phenylenediamines or o-aminobenzothiols with various carboxylic acids in $P_4O_{10}/MeSO_3H$ [507, 508] (Scheme 197).

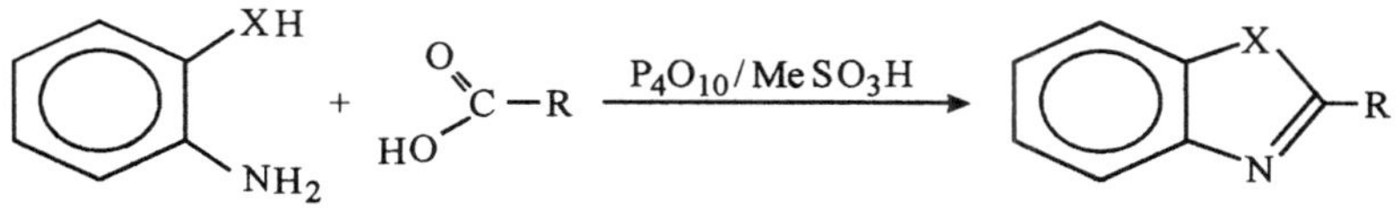

$$30 - 97\% \tag{197}$$

$$X = NH \; ; \; R = Ph, \; p\text{-}Cl\,C_6H_4, \; o\text{-}Cl\,C_6H_4, \; p\text{-}O_2N\,C_6H_4, \; p\text{-}Me\,C_6H_4,$$
$$o\text{-}HO\,C_6H_4, \; m\text{-}Me\,C_6H_4, \; n\text{-}Bu, \; n\text{-}C_5H_{11}, \; n\text{-}C_6H_{13}, \; n\text{-}C_7H_{15}$$
$$X = S \; ; \; R = Ph, \; p\text{-}ClC_6H_4, \; p\text{-}MeOC_6H_4, \; p\text{-}O_2NC_6H_4$$

Only in two cases ($X=NH$; $R=o\text{-}HOC_6H_4$, $p\text{-}O_2NC_6H_4$) were the yields of the products less than 85% [507, 508].

The electron-withdrawing effect in the aromatic acids retarded the process of condensation with o-phenylenediamine. For example, 2,2′-m-phenylene bis[benzimidazole] was isolated with rather low yield (29%) [507] (Scheme 198).

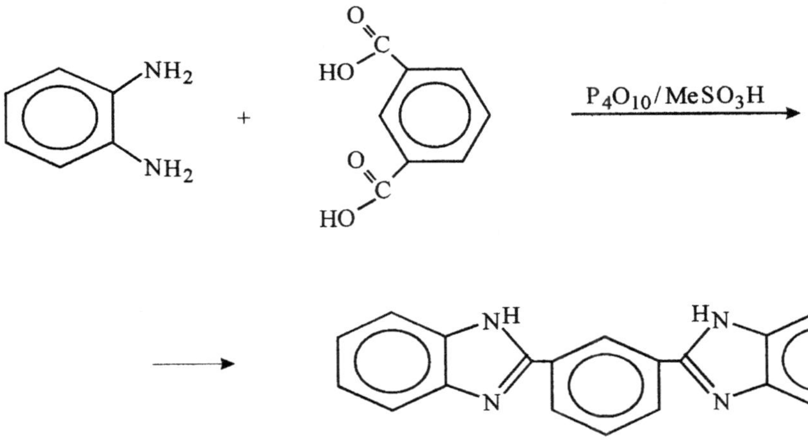

$$\tag{198}$$

$$29\%$$

Nevertheless, 1,2,4,5-tetraminobenzene dihydrochloride underwent condensation with 2 moles of carboxylic acid initiated by $P_4O_{10}/MeSO_3H$ [507] (Scheme 199).

$$\text{(199)}$$

$$91\text{--}94\ \%$$

The polycondensation of 3,3′-diaminobenzidine tetrachloride monohydrate with dicarboxylic acids in $P_4O_{10}/MeSO_3H$ [507] is shown in Scheme 200.

$$\text{(200)}$$

$$R = \text{- Alk -, - Ar -} \qquad n = 5\text{ - }8,\ 10$$

The optimum temperature for the reaction was found to be 140°C and the time of the process could vary from 1.6 to 10 h [507].

Under the action of $P_4O_{10}/MeSO_3H$ 2,5-diamino-1,4-benzenedithiol reacted with 2 equivalents of benzoic acid or cyclohexane carboxylic acid to yield 2,6-diphenyl- or 2,6-dicyclohexyl-bisthiazole in accordance with Scheme 201 [507].

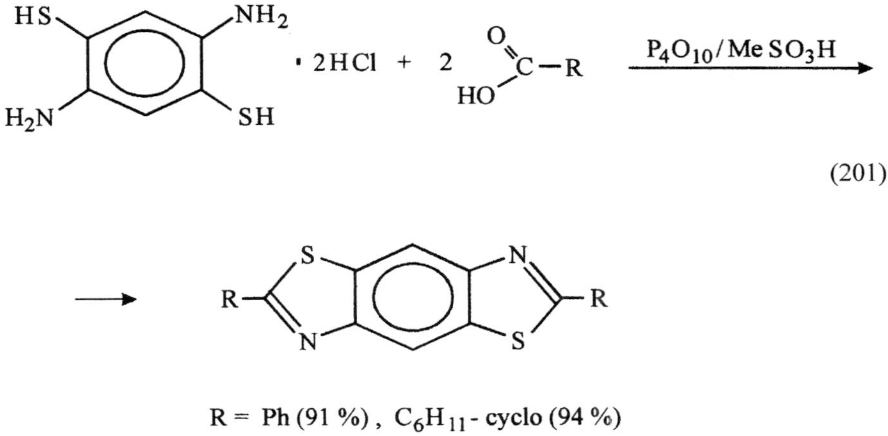

$$(201)$$

R = Ph (91 %), C_6H_{11}- cyclo (94 %)

The reaction of 2,5-diamino-1,4-benzenedithiol with dicaboxylic acids resulted in the products of polycondensation [508].

o-Aminophenol reacted with benzoic acid and its 4-substituted derivatives in P_4O_{10}/MeSO$_3$H to give a mixture of the corresponding benzoxazole and its benzoylated derivative (4- or 5-benzoyl-2-phenylbenzoxazole) [509] (Scheme 202).

o-Aminophenols substituted in the aromatic ring reacted with the same acids with the formation of only benzoxazoles (yields 73–99%).

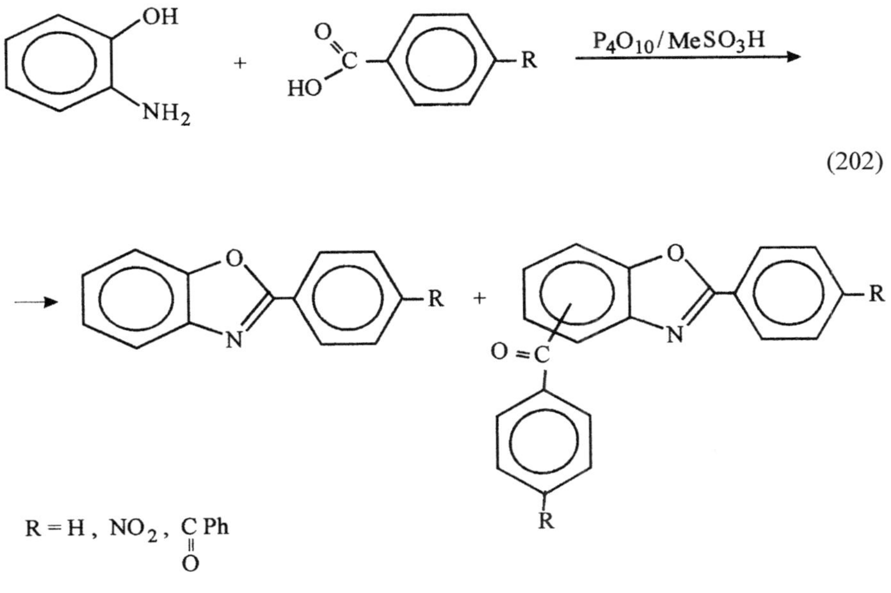

$$(202)$$

R = H, NO$_2$, C Ph ($\|$ O)

o-Aminophenols like o-phenylenediamine and o-aminobenzothiol gave the products of polycondensation in the reactions with dicarboxylic acids [509]. The polymers produced [507–509] were characterized by means of spectroscopic and thermographic methods.

The mixture of phosphoric anhydride with methanesulphonic acid promoted acylation of dibenzocrown ether under very mild conditions [510, 510a].

Dibenzo-18-crown-6 ether underwent direct polycondensation with aliphatic or alicyclic dicarboxylic acids [511] in a homogenous solution of $P_4O_{10}/$ $MeSO_3H$ at 50°C to give high molecular weight poly(crown etherketone) (Scheme 203).

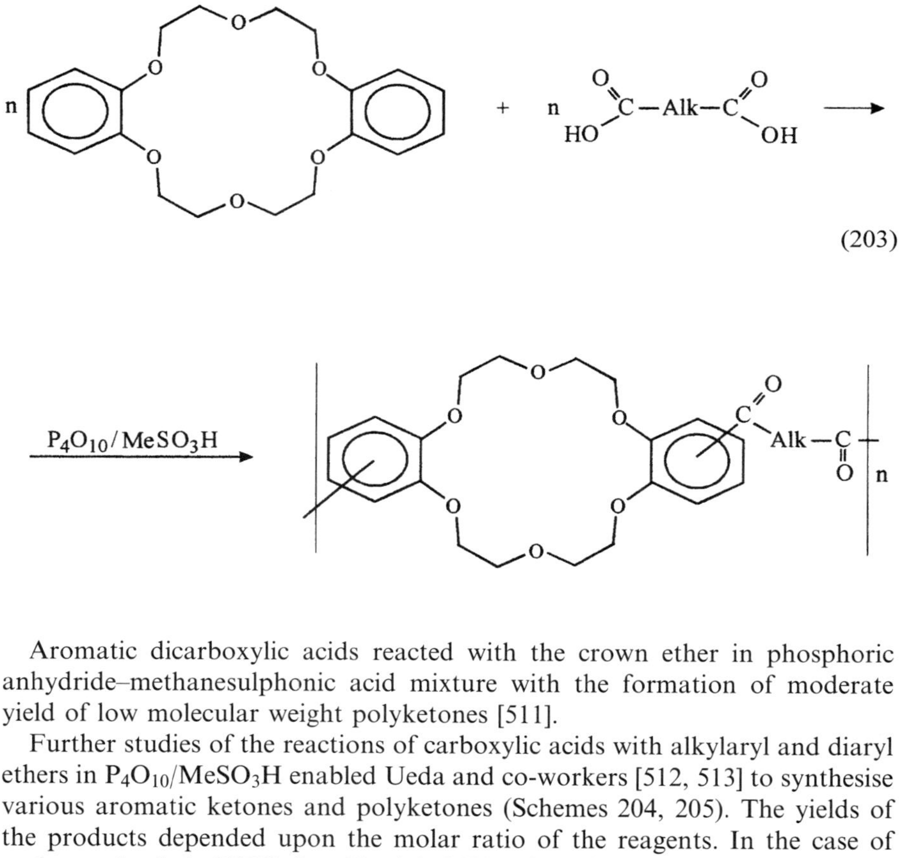

$$(203)$$

Aromatic dicarboxylic acids reacted with the crown ether in phosphoric anhydride–methanesulphonic acid mixture with the formation of moderate yield of low molecular weight polyketones [511].

Further studies of the reactions of carboxylic acids with alkylaryl and diaryl ethers in $P_4O_{10}/MeSO_3H$ enabled Ueda and co-workers [512, 513] to synthesise various aromatic ketones and polyketones (Schemes 204, 205). The yields of the products depended upon the molar ratio of the reagents. In the case of molar ratio 1:1 (60°C for 30 min) 95% of 4-phenoxybenzophenone was isolated. The molar ratio diphenyl ether/benzoic acid 1:2 and prolonged heating (6 h) at the same temperature led to the formation of 4,4′-oxydibenzophenone (95%) [512] (Scheme 204). A comparative study of the reactivity of diphenyl ether and 1,4-diphenoxybenzene in $P_4O_{10}/MeSO_3H$ [513]

revealed the influence of the structure upon the process of dehydration–condensation. Due to the deactivation effect of the benzoyl group dibenzoylation was retarded.

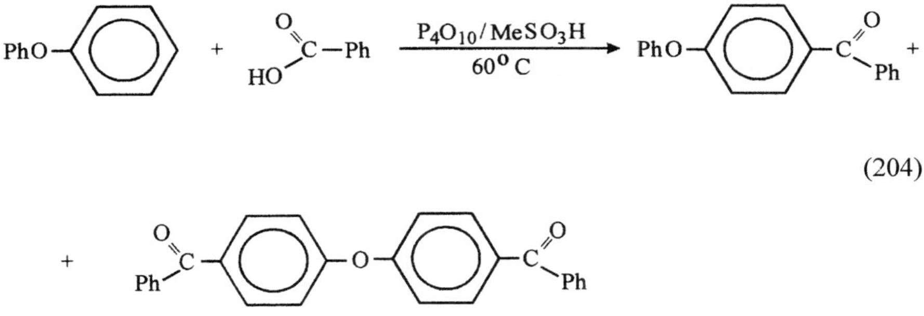

$$(204)$$

Polycondensation of dibenzoic acids with 1,4-diphenoxybenzene in P$_4$O$_{10}$/MeSO$_3$H at 120°C produced high molecular weight poly (etherketones) in a quantitative yield [513] (Scheme 205).

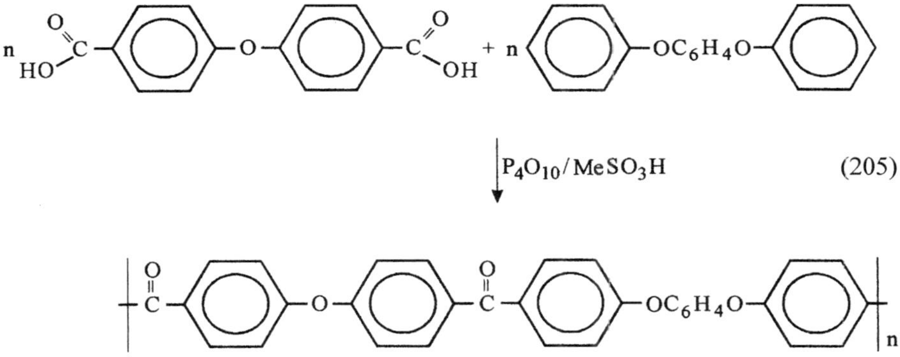

$$(205)$$

The phosphoric anhydride–methanesulphonic acid mixture at 100°C (24 h) also affected the self-condensation of phenoxybenzoic acid in accordance with Scheme 206 [513, 514].

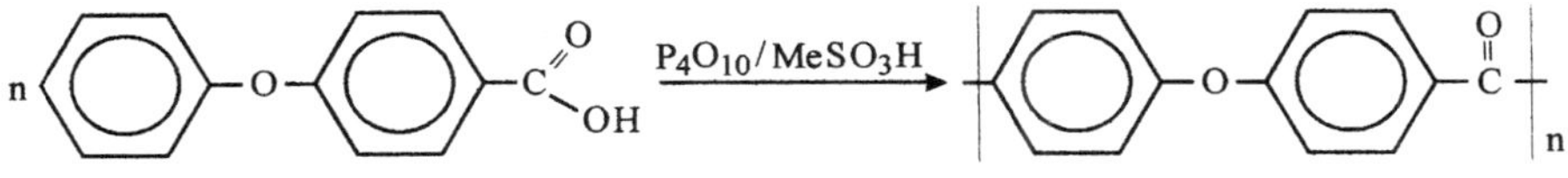

$$(206)$$

The condensation reaction between benzoic acid and anthraquinone as well as the copolymerization of the latter with dicarboxylic acids was carried out in $P_4O_{10}/MeSO_3H$ [515].

A recent example of an intermolecular dehydration–condensation reaction under the action of P_4O_{10} in combination with $MeSO_3H$ is described by Yonezawa and co-workers [516]. The $P_4O_{10}/MeSO_3H$ mixture is said to be among the most active reagents for a direct condensation of carboxylic acids with aromatic compounds resulting in formation of phenones [516] (Scheme 207).

$$\underset{\substack{\\ OH}}{Me-CH-C}\overset{\substack{R \\ |}}{}\overset{O}{\diagup} + HAr \xrightarrow[-H_2O]{P_4O_{10}/Me\,S\,O_3H} \underset{\substack{\\ Ar}}{Me-CH-C}\overset{\substack{R \\ |}}{}\overset{O}{\diagup}$$

$$R = H, Et, Ph, Cl \tag{207}$$

The process was carried out at low temperature (ice bath) and resulted in variable yields of products. There is speculation on the mechanism of the reaction which surprisingly includes reduction of phosphorus atoms with the formation of P–H⋯OOC and P–O–H⋯Ar bonds along with the transformation of the P=O bond into the P–O–H bond. No ^{31}P NMR data are presented in the publication [516].

Pyrroloazepin dione and indanone were also synthesized from the same initial organic compounds under the action of polyphosphoric acid. In the latter case the yields were only slightly (from 5 to 10%) lower than in the reactions with the mixture of phosphoric anhydride with methanesulphonic acid. A combination phosphoric anhydride–trifluoromethanesulphonic acid promoted the reaction with significantly lower yield than P_4O_{10}–$MeSO_3H$ (71% and 90%, respectively). There are no comments about possible causes for the differences. The authors [516] consider that the method developed in their research can find large-scale industrial application due to high yields and a straightforward procedure.

Polyphosphoric acid is an inevitable product formed in the course of dehydration by phosphoric anhydride. It means that only the primary step of such processes takes place entirely under the action of phosphoric anhydride whereas the following steps proceed either under indirect or, more likely, direct influence of polyphosphoric acid, the composition of which changes in the course of dehydration. In this respect the studies of polyphosphoric acid reactions with various organic compounds simulate to a certain extent the later stages of phosphoric anhydride interactions with the same substrates, particularly nucleophilic reagents possessing O-H, S-H and some other functional groups. For obvious reasons polyphosphoric acid displays a number of peculiarities that make it a specific reagent in organic and bioorganic synthesis. The analysis of similarities and dissimilarities between the processes involving phosphoric anhydride and polyphosphoric acid might help to understand their reaction pathways. From the technological point of view in many cases polyphosphoric acid is a more welcoming reagent than phosphoric anhydride due to its significantly lower hydrophilicity.

For polyphosphoric acid the intermolacular dehydrating action is more characteristic than for phosphoric anhydride which readily acts as intra-molecular dehydration–condensation promoter [517, 518]. For example, application of polyphosphoric acid to the synthesis of phenyl esters from carboxylic acids and phenol [519] became a good alternative to the earlier methods that involved phosphorus oxychloride, acid chlorides or acid anhydrides including trifluoroacetic anhydride. A simple procedure gave high yields (85–98%) of phenyl esters (Scheme 208) even with carboxylic acids of relatively low reactivity such as benzoic, diglycolic, levulinic, maleic, methacrylic, phthalic, salicylic and stearic acids.

$$RC\underset{OH}{\overset{O}{\diagup}} \quad + \quad HO-\bigcirc \quad \xrightarrow[-H_2O]{PPA} \quad RC\underset{O}{\overset{O}{\diagup}}-\bigcirc \qquad (208)$$

Klibansky and Ginsburg [520] heated ketoacid with polyphosphoric acid to convert it into the polycyclic diketone (Scheme 209). The same ketoacid was dehydrated by hydrogen fluoride to give a different product. Polyphosphoric acid caused cyclisation into the position 8 in the naphthalene fragment, whereas hydrogen fluoride caused cycloaddition at position 2. It is possible to avoid the use of highly viscous neat polyphosphoric acid in some reactions. Thus the action of polyphosphoric acid (15%) in glacial acetic acid for 10 min in a hot-water bath was sufficient to complete the cyclization process (Scheme 210) with a 82% yield of cyclized product [521].

Some relevant examples of dehydration by phosphoric anhydride, poly-phosphoric acid, polyphosphoric ester and their derivatives are reviewed in Chapter 5.

8.2 DEALCOHOLYSIS

As mentioned above in Sections 3.1.6 and 3.8.1 the first study of diethyl malonic ester dealcoholysis under the action of phosphoric anhydride was described by Diels and Wolf in 1906 [162]. The reaction was performed at 300°C with a thirty fold excess of P_4O_{10}. The evolved gaseous carbon suboxide was condensed in a trap cooled to $-70°C$ (Scheme 211).

$$(209)$$

$$\text{PPA} , \quad \text{Me COOH} \longrightarrow \qquad 82\,\% \qquad (210)$$

$$\underset{\text{COOEt}}{\overset{\text{COOEt}}{H_2C}} \xrightarrow[-\text{EtOH}]{P_4O_{10}} \quad O=C=C=C=O \qquad (211)$$

This attractive method of malonic ester dealcoholysis was expanded to other esters on a limited scale, which seems strange because the products of the process (ketenes) are of theoretical and practical importance as reaction intermediates in many chemical processes and acylating agents [522]. The latter property of ketenes has been used successfully in modern industry.

In 1978 Eleev and co-authors [523] patented a synthetic method for fluorosulphonyltrifluoromethylketene which involved treating alkyl esters of

2-fluorosulphonyl-3,3,3-trifluoropropionic acid with phosphoric anhydride (Scheme 212). The reaction was initiated by rapid heating of the reagent mixture to 100–110°C after which the temperature was raised to the temperature of the ketene boiling point (160°C). The maximum yield was achieved for the iso-propyl ester. The structure of the ketene was correlated with two characteristic bands in IR spectra: 1780 cm^{-1} (C=O) and 2200 cm^{-1} (C=C=O) [523].

$$F_3C\diagdown CH-C\diagup^{O}_{\diagdown OAlk} \quad \xrightarrow[-\,AlkOH]{P_4O_{10}} \quad F_3C\diagdown C=C=O \qquad (212)$$

42 - 68 %

The action of phosphoric anhydride upon the monosubstituted acetic esters was studied by the authors of this book [163–165, 215, 524–526]. Reactions of esters containing electron-withdrawing groups (Cl and NO$_2$) started at 70–80°C and proceeded exothermically. The reactions of the alkyl esters of phenylacetic, cyanoacetic and malonic acids with P$_4$O$_{10}$ required continuous external heating. The general reaction pathway was dealcoholysis of the esters leading to the formation of monosubstituted ketenes that were reactive under the process conditions [163, 164, 215, 524] (Scheme 213).

Interaction between P$_4$O$_{10}$ and esters proceeded probably via electrophilic addition of P$_4$O$_{10}$ to an ester followed by elimination of the alkoxyphosphoryl group. The formation of acetyl phosphoric and acetyl pyrophosphoric esters [163, 164] could take place in the reaction of monosubstituted ketenes with polyphosphoric acid [163] (Scheme 214) similarly to the process of phosphorylation of nonsubstituted ketene described in the publication [257] (Scheme 215).

It is most likely that arylphosphates were formed by the phosphorylation of accumulated phloroglucinols by either unreacted phosphoric anhydride or polyphosphoric acid. Phosphorylation of the reaction intermediates followed by their cyclotrimerization was ruled out because of the formation of not only triphosphorylated derivatives but also of mono and diphosphorylated phloroglucinols as well [163, 164].

Analysis of the solid residue of the reaction between phosphoric anhydride and diethyl malonic ester (Scheme 216) resulted in the isolation of a number of phosphorylated polysubstituted benzenes. They were 2,4,6-tricarboxy-1-dihydroxyphosphoryl phloroglucinol, 2,4,6-tricarboxy-1,3,5-tris (tri-hydroxyphosphoryl) phloroglucinol and bis (2,4,6-tricarboxy-3,5-dihydroxyphenyl) phosphoric acid [163, 164, 215]. The products can be considered to be the result of carbon suboxide transformations under the action of an excess of phosphoric anhydride and polyphosphoric acid at high temperature. The dominating transformations of ketenes and carbon suboxide was cyclotrimerization. Some of these compounds are described in Section 3.1.6.

$$R = Alk$$
$$Z = Ar, Cl, CN, NO_2$$
$$\textcircled{P} = PO_3H, P_2O_6H_3$$

(213)

The isolation of cyanoketene was reported by Efremov and Zavlin [163, 524]. It was condensed as a colourless liquid with the boiling point $-34°C$ and could be stored for a short period of time in inert solvents at low temperature. The only proton was characterized by the chemical shift 5.28 ppm (in diethyl ether) which was very close to the chemical shift of the proton in fluoroalkylketene [448–451]. ^{13}C NMR spectra of cyanoketene contained three broad signals in the regions (δ, ppm) 52.2 (C–H), 118.4 (CN), 208.7 (C=O) which is consistent with the corresponding data of the earlier publication [527] taking in consideration the electron withdrawing influence of the cyano group.

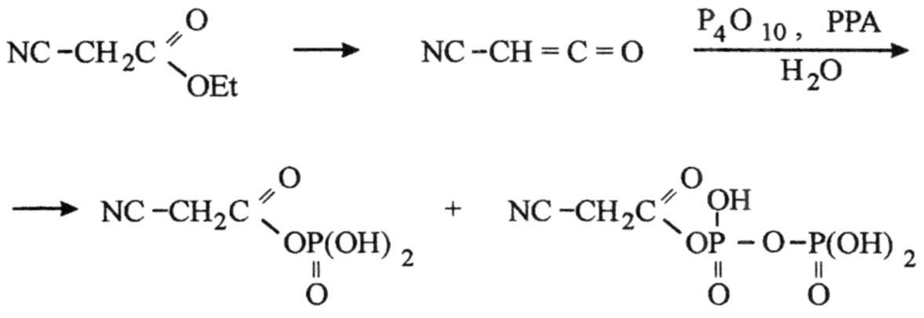

$$NC-CH_2C\underset{OEt}{\overset{O}{\diagup}} \longrightarrow NC-CH=C=O \xrightarrow[H_2O]{P_4O_{10},\ PPA}$$

(214)

$$\longrightarrow NC-CH_2C\underset{\underset{O}{\overset{\|}{OP(OH)_2}}}{\overset{O}{\diagup}} + NC-CH_2C\underset{\underset{O}{\overset{\|}{OP}}-O-\underset{O}{\overset{\|}{P(OH)_2}}}{\overset{O}{\diagup}\overset{OH}{\underset{|}{}}}$$

PPA = Polyphosphoric acid

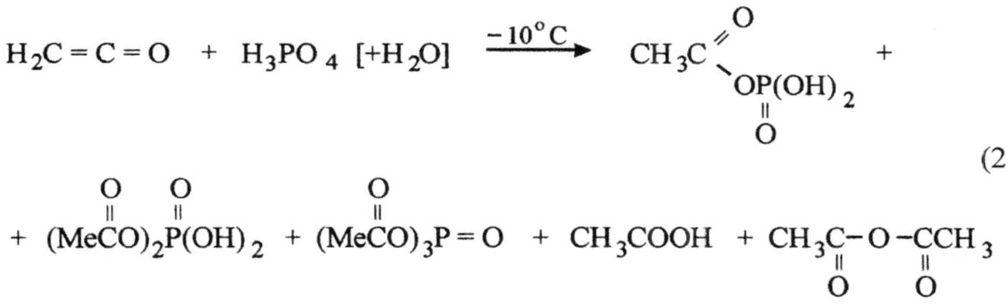

$$H_2C=C=O + H_3PO_4\ [+H_2O] \xrightarrow{-10°C} CH_3C\underset{\underset{O}{\overset{\|}{OP(OH)_2}}}{\overset{O}{\diagup}} +$$

(215)

$$+ (MeCO)_2P(OH)_2 + (MeCO)_3P=O + CH_3COOH + CH_3\underset{O}{\overset{\|}{C}}-O-\underset{O}{\overset{\|}{C}}CH_3$$

$$EtOOCCH_2COOEt \xrightarrow[-EtOH]{P_4O_{10}} O=C=C=C=O \xrightarrow[2.\ H_2O]{1.\ P_4O_{10}}$$

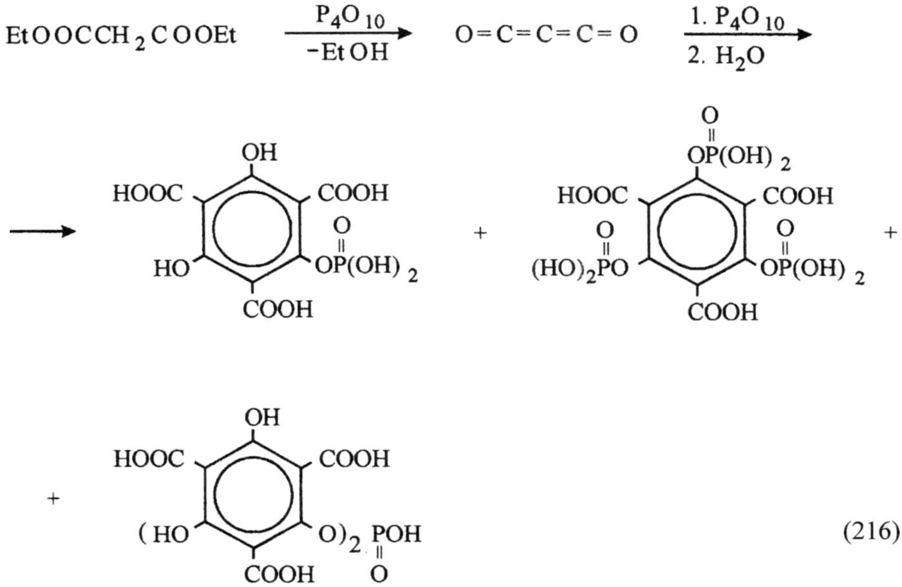

(216)

Dealcoholysis of monosubstituted acetic esters under the action of phosphoric anhydride did not come as a surprise because elimination of alcohols from malonic esters by P_4O_{10} has been known since the beginning of this century [521] (Sections 3.1.6 and 3.8.1). Disubstituted cyanoketenes had been accumulated as reactive intermediates by Moore and co-workers [528–531] in the thermolysis of aza compounds in accordance with Schemes 217 and 218.

$$\text{(217)}$$

$$\text{(218)}$$

R = Alk, Hal

The ketenes were usually reacted further *in situ* predominantly in the subsequent reactions of cycloaddition [440, 532, 533] though some substituted ketenes particularly allenylketenes [534] have been found to be relatively stable compounds. Cycloaddition is characteristic for monosubstituted ketenes as well. One of the recent examples is cycloadditions of silylketenes to 1,3-dienes [535] and isocyanates [536, 537, 538]. The latter process was studied in depth experimentally [536, 537] and theoretically [538].

Cyanoketene was generated by Bock and co-workers [539] in pyrolysis of β-cyanoacetyl chloride (Scheme 219).

$$\text{NC C H}_2\text{C}\overset{O}{\underset{Cl}{\diagdown}} \xrightarrow{-HCl} \text{NC C H} = \text{C} = \text{O} \qquad \text{(219)}$$

In 1997 C. Wentrup and co-workers reported [539b] the generation of cyanoketene by flash vacuum thermolysis of various derivatives of cyanoacetic acid or Meldrum's acid. The vapour was condensed at 14 K. The identity of the product was supported by its mass spectrum and comparison of experimental and calculated IR spectral bands at 2163 and $2239\,\text{cm}^{-1}$. These authors reported that pure cyanoketene was very reactive and started to react (presumably with itself) when the temperature was raised to 80 K.

Cyanoketene evolved from the interaction of cyanoacetic ester with phosphoric anhydride [524] reacted readily with water, alkanols and amines to give cyanoacetic acid, its corresponding esters and amides [163, 164, 215] in accordance with Scheme 220.

Chloroketene [163, 164] and nitroketene [163, 164, 526] were condensed as heavy colourless liquids in traps cooled by liquid nitrogen. Warming the traps caused ketene transformation into viscous products of polymerization. Chloroketene and nitroketene reacted readily with water and alcohols [163, 164, 526] with the formation of corresponding acids or esters. Titration of chloroacetic, cyanoacetic and nitroacetic acids derived from the corresponding ketenes made it possible to determine the yields of chloroketene (17%), cyanoketene (33%) and nitroketene (15%). Phenylketene and 2,4-dichlorophenylketene did not distil from the reaction mixtures because of their very low vapour pressure and high reactivity under the actual reaction conditions.

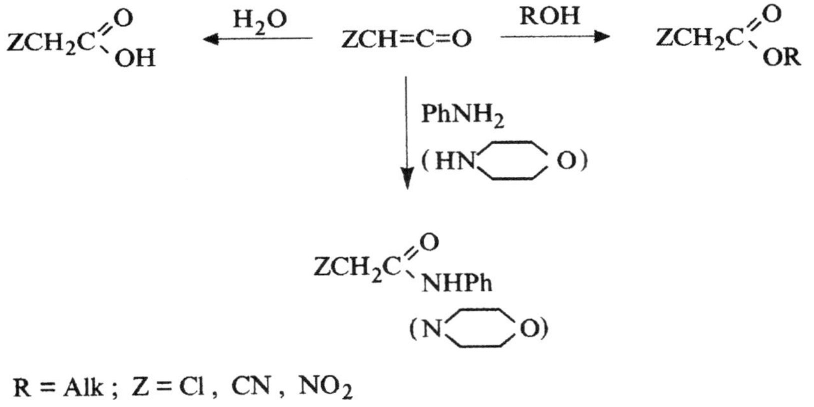

$$R = \text{Alk}; \ Z = \text{Cl}, \ \text{CN}, \ \text{NO}_2 \tag{220}$$

Disubstituted ketenes were shown to undergo cyclotrimerization during the 1960s [440, 447, 540, 541]. Dimethylketene and diphenylketene trimerized to give hexamethylcyclohexanetrione and hexaphenylcyclohexanetrione respectively under the action of strong base. Tetramethyl(aryl) cyclobutanediones were considered to be the intermediates of the cyclotrimerization process as presented in Scheme 221.

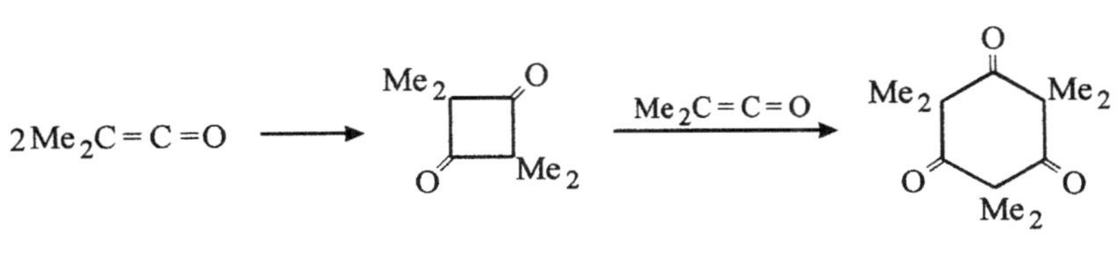

$$\tag{221}$$

The authors of the latter publication [540] reduced the hexamethylcyclohexanetrione to 2,2,4,4,6,6-hexamethylcyclohexanetriol [540] (Scheme 222).

$$(222)$$

As a consequence of the structural features hexamethylcyclohexanetrione could not isomerize into the corresponding phloroglucinol unlike trisubstituted cyclohexanetriones derived from the corresponding monosubstituted ketenes (Scheme 223).

$$(223)$$

Though methods for the synthesis of phloroglucinols have been studied for more than a century [542–544] the reaction of esters with phosphoric anhydride probably presents the most straightforward approach to their substituted derivatives.

Besides phloroglucinols the reactions of monosubstituted esters with phosphoric anhydride gave mixed esters of acetic and phosphoric or pyrophosphoric acids, disubstituted cyclobutane-1,3-dione and 2,4,6-trisub-stituted cyclohexane-1,3,5-trione [163, 164, 215]. The formation of these compounds presented some evidence on possible mechanisms of monosub-stituted ketene cyclotrimerization [163]. The mechanism proposed involved reaction of the ketene with its ynol tautomer since concerted cycloaddition reactions are most favoured when the energies of the LUMO of one reactant is close to the HOMO of the other and this is best achieved by having an 'electron deficient' molecule reacting with an 'electron rich' molecule. This gave an intermediate stabilised by conjugation which can react with a further molecule of ketene to give, after ring opening, a hexadienone. Tautomerisation would give the final phloroglucinol product. A rather unorthodox alternative mechanism suggested for the process [163] involved the construction of a six-membered ring on a rigid tricyclic P_4O_{10} matrix which was supported by molecular mechanics docking calculation. Another mechanism was based upon the established reversible rearrangement of ketenes into the hydroxyacetylenic structures that could be the subject of O-phosphorylation by excess phosphoric anhydride or polyphosphoric acids. This could be followed by the process of dimerization (Scheme 224).

$$2 \ Z-CH=C=O \longrightarrow \quad \xrightarrow{\ Z\,CH=C=O\ } \quad \tag{224}$$

$$2 \ Z-C\equiv C-OH \longrightarrow \quad \xrightarrow{\ Z-C\equiv C-OH\ }$$

The probability that the reaction follows this mechanism was supported by the isolation of 2,4-diphenylcyclobutane-1,3-dione in a yield of 26% [163]. It is natural to expect the further insertion of another molecule of acetylene phosphate into the dimer at the elevated temperature of the process (Scheme 224). This reaction pathway leads to the formation of the phloroglucinol triphosphates or tripyrophosphates that were isolated as the major products of phenylketene cyclotrimerization (total yield 67%) but this is not the case with cyanoketene cyclotrimerization which gave only 26% yield of tricyanophloroglucinol triphosphate and tripyrophosphate [163, 164].

The isolation of mixed esters of acetic and phosphoric or pyrophosphoric acids and cyclohexanetrione (Scheme 213) presented some evidence for a mechanism involving ketene phosphorylation (nucleophilic addition) with the formation of the carbanion 'a' followed by the two steps addition of the ketene to carbanions 'b' and 'c' as it is shown in the Scheme 225. Whether P$_4$O$_{10}$ is sufficiently nucleophilic to generate carbanionic intermediates is open to debate.

So, Efremov and Zavlin discovered a new type of monosubstituted ketenes transformation — cyclotrimerization leading directly to the polysubstituted aromatic systems.

Another example of cyclotrimerization initiated by phosphoric anhydride was presented, i.e. α-sulphofluoro tetrafluoropropionic nitrile high yield condensation into the corresponding trimer (Section 3.8.1, Scheme 181).

The synthesis of various highly substituted aromatic compounds from the corresponding structural blocks is of pronounced synthetic significance. The procedure has a number of advantages [545, 546] over widely used electrophilic substitution of arenes. For instance, it is possible to synthesize directly aryl compounds possessing substituents in orientations that contradict the general directing rules governing electrophilic substitution. The synthesis of highly substituted benzenes from noncyclic precursors is a rapidly developing branch of organic chemistry [545, 546].

$$\text{ZCH}=\text{C}=\text{O} \xrightarrow{\ P_4O_{10}\ } \underset{a}{\text{Z}\overset{\ominus}{\text{C}}\text{HC}\underset{\text{O}}{\overset{\text{O}-\text{P}}{\diagup}}} \xrightarrow{\ \text{ZCH}=\text{C}=\text{O}\ }$$

(225)

$$\xrightarrow{\quad} \underset{b}{\underset{\text{O}=\text{C}-\overset{\ominus}{\text{C}}\text{HZ}}{\overset{\text{ZCH}-\text{C}\diagup^{\text{O}-\text{P}}_{\text{O}}}{\Big|}}} \xrightarrow{\ \text{ZCH}=\text{C}=\text{O}\ } \underset{c}{\bigcirc} \xrightarrow{\quad}$$

$$\xrightarrow{\quad} \text{(cyclohexane-1,3,5-trione ring, substituents H, Z)}$$

Shvedova and co-authors [458] performed intramolecular dealkoxylation–cyclization using phosphoric anhydride. Dealcoholysis of perfluoro glutaric amidoester occurred at high temperature to give a cyclic amide of perfluoroglutaric acid (Scheme 226).

$$\text{EtOOC}(\text{CF}_2)_3\underset{\text{NH}_2}{\overset{\text{C}=\text{O}}{\underset{|}{}}} \xrightarrow[-\text{EtOH}]{\ P_4O_{10}\ } \text{O}=\text{C}\underset{\text{NH}}{\overset{(\text{CF}_2)_3}{\diagup\diagdown}}\text{C}=\text{O}$$

(226)

Cho and co-authors [547, 548] carried out intramolecular dealcoholysis–cyclization for substituted pyrroles in accordance with Scheme 227. In this case phosphoric anhydride in either methanesulphonic acid or trifluoromethane-sulphonic acid was employed as the reagent with no preference of one or another reagent.

The reactions were complete within 3–10 min at 80–100°C with yields of 71–95% [548]. This method does not have the disadvantages of the highly viscous polyphosphoric acid which is said to require up to 30-fold excess of a reagent.

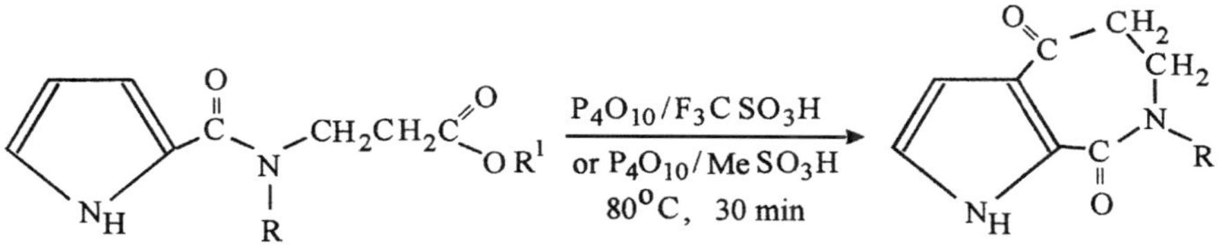

$$R = Alk, \ Alk\text{-}Ar; \quad R^1 = Alk, \ Alk\text{-}Ar, \ Ar \tag{227}$$

Phosphoric anhydride in methanesulphonic acid induces the very interesting, but complicated, reaction shown in Scheme 228. Diarylation occurs with decarboxylation and the elimination of a methoxyl group. Yonezawa and co-workers [516] state that the reaction proceeds very smoothly.

$$MeO\text{-}CH\text{-}C \overset{O}{\underset{OH}{}} + \ HAr \ \xrightarrow{P_4O_{10} / MeSO_3H} \ Me\text{-}CHAr_2 \tag{228}$$

$$R = H, \ Et$$

A rare case of dealcoholysis under the action of polyphosphoric acid was published by Walker [549] in one of his impressive papers devoted to PPA. The substituted ethyl β-hydroxy propionate was dealcoholylated and dehydrated by polyphosphoric acid into an indenone (Scheme 229).

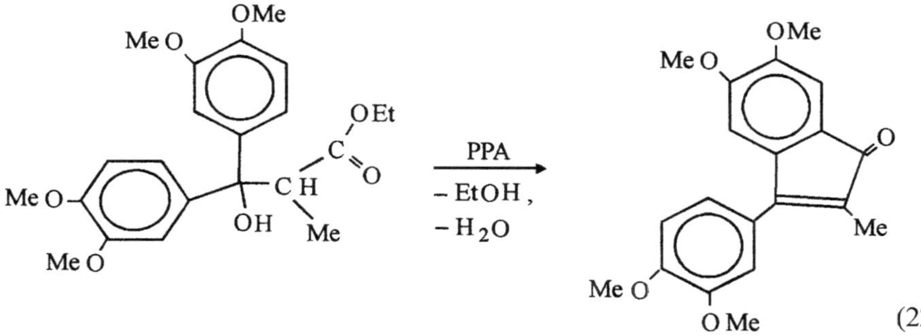

$$\xrightarrow[-H_2O]{\substack{PPA \\ -EtOH,}} \tag{229}$$

But the above reaction illustrates the exception rather than the rule. For example Walker and co-authors [550] reacted 2,3-dicarboxy propionate with polyphosphoric acid to give entirely the product of dehydration (Scheme 230) without any signs of dealcoholysis.

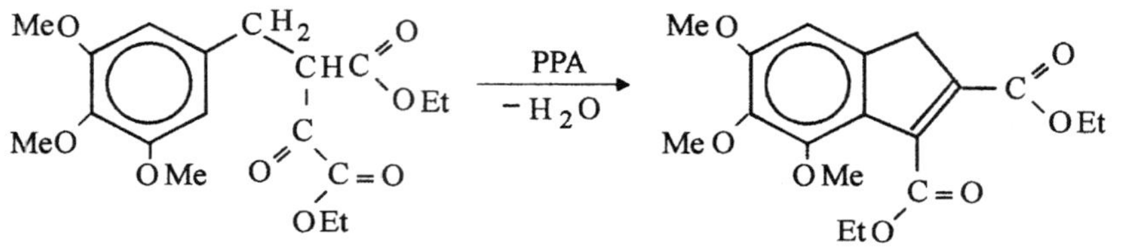

92 % (230)

8.3 OTHER ELIMINATION REACTIONS

Phosphoric anhydride caused the elimination of ethylene and sulphur dioxide from ethyl 2-ketopentafluoropropanesulphonate at 125°C leading to 2-ketopentafluoropropyl 2-ketopentafluoropropanesulphonate [551] in accordance with Scheme 231.

$$CF_3\underset{\overset{\|}{O}}{C}CF_2SO_2OEt \xrightarrow[-CH_2=CH_2,\ -SO_2]{P_4O_{10}} CF_3\underset{\overset{\|}{O}}{C}CF_2SO_2OCF_2\underset{\overset{\|}{O}}{C}CF_3$$

28% (231)

Phosphoric anhydride was also involved in the process of elimination of the alkylthio group in bioorganic synthesis [552]. This reaction is discussed in Chapter 5.

8.4 CONDENSATION, REARRANGEMENT AND SUBSTITUTION

Phosphoric anhydride can act as a catalyst for some rearrangements. In several publications devoted to this property of P_4O_{10} its role is described as catalytic though from the classical point of view such a term is not correct because small portions of phosphoric anhydride can take an active part in the processes but eventually it does not convert reversibly to its initial form.

Acetylcyclohexane was prepared by Saunders [553] by refluxing the mixture of 1-ethynylcyclohexanol with phosphoric anhydride in benzene for 2.5 h Scheme 232.

56 - 70 % (232)

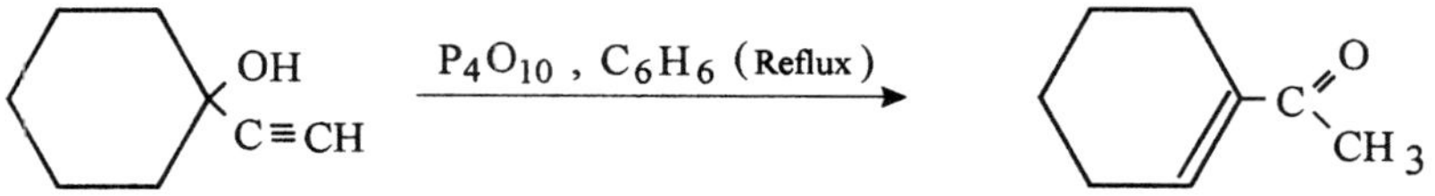

The process of cyclization–rearrangement of alkylstyrenes and related reactions under the action of phosphoric anhydride were studied by Condon and co-workers [554, 555] and other scientists [556, 556a, 557] as a refinement of studies made many years earlier [557a]. The substituted alkylstyrenes synthesized in the Grignard reaction of benzaldehydes and corresponding alkylmagnesium bromides were dehydrated by either small amount of p-toluenesulphonic acid or 5% by weight of phosphoric anhydride. The latter dehydrating agent appeared to be more efficient because it did not initiate the formation of the corresponding substituted dibenzyl ethers as monitored by NMR spectrometry [554]. After the addition of a fresh portion of P_4O_{10} and refluxing of the reaction mixture substituted 1,2,3,4-tetrahydronaphthalenes was formed in high yield as shown in Scheme 233 [554, 555]. In some cases a minor quantity (ca 9%) of 1-ethyl indane and traces of the polymeric product were isolated [554]. Presumably the P_4O_{10} catalyses the migration of the double bond 5,6 position before activating it for the cyclisation step. The product structures were confirmed, along with spectrometric methods, by their dehydrogenation under the action of 10% palladium–charcoal catalyst into the corresponding substituted naphthalenes (Scheme 233).

$$(233)$$

27 - 81 %

R = H , Me

As pointed out by the authors [554, 555], this new type of reaction could be of considerable synthetic importance particularly for preparation of indanes, tetralins, naphthalenes and their analogues. This was supported by the synthesis of 1-methylphenanthrene with the 23% overall yield starting from 1-bromonaphthalene using the same method [555] (Scheme 234).

Hexafluoroacetone hydrazone was condensed with a variety of ketones under the action of phosphoric anhydride at 20–100°C (1–12 h) in dichloromethane [556]. Unsaturated anhydrides were synthesized in 77% yield by Yu and Kayser [557] from diesters under the influence of phosphoric anhydride *in vacuo* at 160°C (Scheme 235).

Sulphur dioxide adds across one of the double bonds of hexafluorobutadiene upon 18 h irradiation in the presence of P_4O_{10} [558]. Phosphoric anhydride and triflic anhydride converted methyl ketone into its sulphonyl enol ether [559] (Scheme 236).

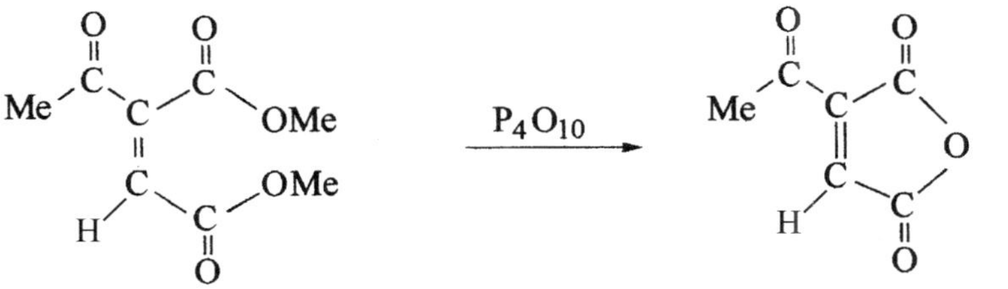

$$\text{(235)}$$

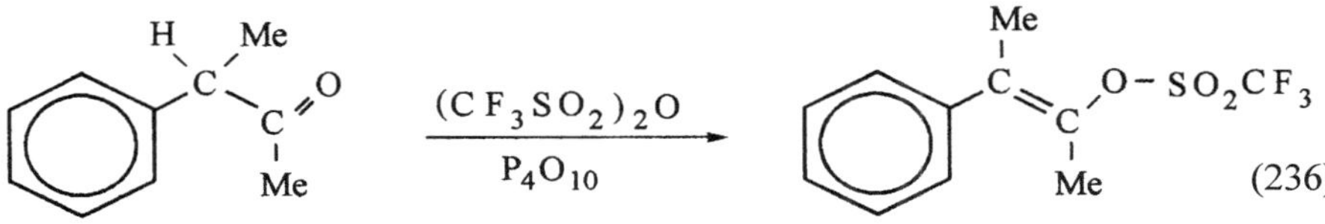

$$\text{(236)}$$

Polyphosphoric acid is also known to induce rearrangement prior to cyclisation. For example according to the patent issued to Bader and Hyre [560] it initiated the rearrangement–condensation of N-butenyl aniline into 2,3-dimethylindole with high yield (Scheme 237).

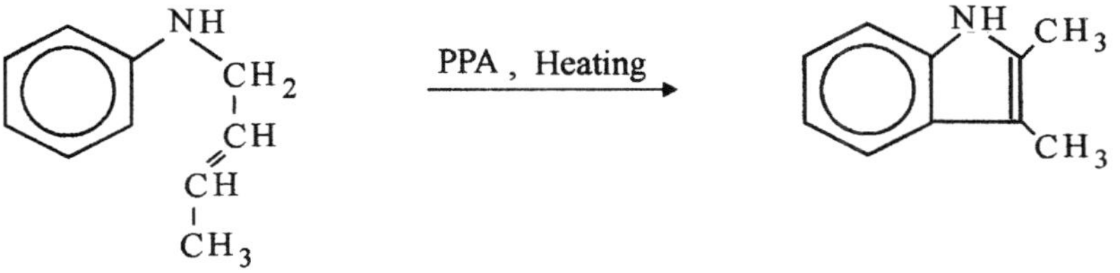

$$\text{(237)}$$

Thesing and Funk [561] used polyphosphoric acid to aid the synthesis of the polycyclic system 4-(1,2,3,4-tetrahydro-1-isoquinolyl) dihydrocarbostyril. The latter compound was prepared by either rearrangement–cyclization of heteroauxin (β-phenethylamine) initiated by polyphosphoric acid (Scheme 238) or in dehydration of the amide derivative by polyphosphoric acid followed by hydrogenation (Scheme 238).

Succinic acid (2 moles) was condensed by the action of polyphosphoric acid with hydrazine hydrate (1 mole) to give disuccinhydrazide in good yield [562] (Scheme 239).

Polyphosphoric acid has also been found to be effective for catalysing the conversion of unreactive ketones to their oximes [563], despite the fact that most oximes undergo the Beckmann rearrangement under similar conditions. The usual hydroxylamine hydrochloride reagent can be replaced by nitromethane which is converted to formic acid and hydroxylamine under the reaction conditions (Scheme 240).

(238)

(239)

(240)

One of the earlier attempts to improve the activity of PPA was the use of sulphuric acid in place of 85% phosphoric acid [564]. The use of 30% $P_4O_{10}\cdot2H_2SO_4$ under very mild conditions (60°C) induced the alkylation of phenol by iso-butanol to give a 95% yield of 4(t-butyl)phenol (Scheme 240a).

$$\text{OH} \quad + \quad \text{Alk}-\text{OH} \quad \xrightarrow{P_4O_{10}/H_2SO_4} \quad \text{OH} \ldots \text{Alk} \tag{240a}$$

The phosphoric anhydride–methanesulphonic acid mixture is also known for promoting rearrangements [565], for example in alkanone oximes (Scheme 241) and condensation as illustrated by the cycloaddition (Scheme 242).

$$C_6H_{13}-CH=N\diagdown_{OH} \quad \xrightarrow[100°C,\ 1\ h]{P_4O_{10}/MeSO_3H} \quad C_6H_{13}-C\diagup^{O}\diagdown_{NH_2} \tag{241}$$

$$\xrightarrow[100°C,\ 1\ h]{P_4O_{10}/MeSO_3H} \tag{242}$$

Polyphosphoric acid in combination with sodium iodide or potassium iodide has been used to convert many primary and secondary alcohols into the corresponding alkyl iodides [566, 567] (Scheme 243).

$$HOCH_2(CH_2)_4CH_2OH + 2\,KI \xrightarrow[\substack{-KH_2PO_4 \\ -H_2O}]{P_4O_{10}/H_3PO_4} ICH_2(CH_2)CH_2I$$

$$83\text{–}85\% \tag{243}$$

Freshly made polyphosphoric acid initiated the preparation of 1,4-dibutyliodide from an ether (tetrahydrofuran) and potassium iodide under refluxing for 3 h in very high yield [568] (Scheme 244).

$$\text{(THF)} + 2\,KI \xrightarrow[\substack{-\,KH_2PO_4 \\ -\,H_2O}]{P_4O_{10}/H_3PO_4} I-CH_2CH_2CH_2CH_2-I \qquad (244)$$

Polyphosphoric acid initiated the iodination of cyclohexene by potassium iodide at the elevated temperature [569] (Scheme 244a).

$$\text{(cyclohexene)} \xrightarrow[P_4O_{10}/H_3PO_4]{KI,\ 80^{\circ}C,\ 3\ hrs} \text{(iodocyclohexene)} \qquad (244a)$$

Polyphosphoric ester (PPE) (Langheld-ester), as described in Section 3.1.3, is made from diethyl ether and phosphoric anhydride. It is widely used as an organic reagent, e.g. for the synthesis of lactones [264]. Its application to nitrogen compounds have been investigated in some depth by Kanoaka [570a–c]. Thus it converts phenyl-substituted benzamides in boiling chloroform solution into their nitriles within 2 h [570a]. It is also favoured for the activation of amino groups for condensation reactions with carboxylic acids. The reagent is believed to convert the amino groups into intermediates with phosphamido groups [570b]. The reagent must be handled with caution since it may contain some highly toxic tetraethyl pyrophosphate. The reagent (5 parts) has also been used for the Bischler–Napieralskii synthesis of 2-substituted benzamidazoles from *o*-phenylenediamines (see above). The reagent [570c] can also replace PPSE in the reaction of carboxylic acids with *o*-aminophenols to produce oxazoles as shown in Scheme 246. In the case of carboxylic acid chlorides, activation can be sufficient to induce cyclisation to an unactivated aryl ring. The reaction conditions are also quite mild, i.e. 70°C for 15 min [245d]. It is an efficient catalyst in the synthesis of mercaptans from hydrocarbon vapour and hydrogen sulphide (Scheme 245) as well as ortho- and poly-phosphoric acids [570e].

$$Alk\,H\,(Vapour) + H_2S \xrightarrow{PPE} Alk\,S\,H \qquad (245)$$

Polyphosphoric acid trimethylsilyl ester (PPSE) is one of the widely used derivatives of phosphoric anhydride and is produced from P_4O_{10} and trimethylsilanol or hexamethyldisiloxane. It acts as an active condensing reagent. For example 2-methylbenzoxazoles, imidazoles and thiazoles have been synthesised from the corresponding o-substituted anilines and acetic acid under the action of PPSE [377] (Scheme 246).

$$\text{(246)}$$

$$\text{(247)}$$

$$X = NH, O, S$$

The Beckmann rearrangement of oximes in polyphosphoric ester [571, 571a] has common features with the reaction of N-acetyl or N-benzoyl o-diaminobenzenes with phosphoric anhydride [215, 376] (Section 3.3.6). PPSE acted in this case as a more active dehydrating-condensing agent than polyphosphoric ester (PPE) [572]. The authors [572] proposed the probable reaction pathways but did not consider the possibility of phosphorylation–dehydrophosphorylation processes.

An interesting approach to application of PPSE was proposed by Yamamoto and Watanabe [377]. The authors prepared the so-called 'heterogenized PPSE' by heating phosphoric anhydride with dry silica gel in 1,2-dichloroethane and used it for the condensation of phenylhydrazine with diethylketone. The resulting 2-ethyl-3-methylindole was isolated in a yield of 88%.

The higher dehydrating activity of PPSE compared with that of phosphoric anhydride has also been exploited. This property of PPSE is based probably upon the high population of branched phosphorus atoms (P_b) in it and the trimethylsilyl groups pronounced tendency to migration. These features help PPSE to activate carbon atoms of the carbonyl group in reactions of condensation. Tris(trimethylsilyl) phosphate was practically inert in such processes [377]. PPSE efficiently dehydrated carboxyamides into nitriles [573]. It is important to emphasize that consistent results for the reactions of PPSE with organic compounds can be achieved only in dry media [102, 377, 410, 571, 571a, 574].

Imamoto and co-authors [410] claimed that a mixture of PPSE with NaI initiated reduction of sulphoxides, α-haloketones and benzoins to the corresponding sulphides, ketones and deoxybenzoins.

The same authors [410] carried out a crossed aldol condensation under the action of PPSE. It was anticipated that PPSE could generate enol-like structures (trimethylsilyl enol and enol phosphate) from the ketone fragments and initiate aldehydes by their interaction with the trimethylsilyl or phosphate group of PPSE. In other words PPSE was anticipated to promote condensation of carbonyl compounds. The process between acetophenone and benzaldehyde was carried out in the presence of PPSE [410]. Eventually instead of the anticipated aldol (1,3-diphenyl-3-dihydroxypropane-1-one), 5-benzoyl-2,4,6-triphenyl-1,3-dioxane was isolated [410, 575] (Scheme 248) the structure of which was determined on the basis of ^{1}H, ^{13}C NMR, IR, mass spectrometry and X-ray data.

$$\text{Ar}\overset{\overset{\displaystyle O}{\|}}{C}\text{Me} \ + \ 3\,\text{Ar'}\,C\overset{\displaystyle O}{\underset{\displaystyle H}{}} \quad \xrightarrow{\ \text{PPSE}\ } \quad$$

(248)

A characteristic feature of the process was the formation of mezo-5-acyl-1,3-dioxane with equatorial orientation of the aryl substituents in positions 2, 4 and 6 and axial orientation of the acyl group. Although one can consider the process to be complicated, it was a one-pot stereoselective synthesis. PPSE was of top importance in this reaction in which tris(trimethylsilyl) phosphate was inert [410]. Introduction of pyrophosphoric esters instead of PPSE in the reaction resulted in an aldol condensation and the formation of α,β-unsaturated carbonyl compounds [410].

Under the action of PPSE 3-(p-chlorophenyl)-3-hydroxy-1-phenylpropane-1-one condensed with benzaldehyde at 0°C to give a mixture of isomers [410] (Scheme 249).

The authors of the paper [410] described the experimental determination of the unusual reaction pathway. Such type of condensation is of pronounced importance for the synthesis of biologically active compounds [576–578] as well as regio and stereoselective synthesis of acyclic products [579].

A rather interesting, albeit surprising, analytical method for aromatic acids is based on their transformation into corresponding aldehydes under the action of polyphosphoric acid (PPA) [580] (Scheme 250) prepared by hydration of phosphoric anhydride (Section 1.1).

$$PhCCH_2CHR + 2PhC \xrightarrow{PPSE} \quad (249)$$

$$R = Alk, Ar$$

$$HOOC\text{-}Ar\text{-}COOH \xrightarrow[\text{Heating}]{PPA} OHC\text{-}Ar\text{-}CHO \quad (250)$$

CONCLUSIONS TO THE CHAPTER

The analysis of the literature data and our experiments demonstrate the versatility of phosphoric anhydride as a phosphorylating agent. One of its specific features is its adamantan-like structure that incorporates four similar (from the stereochemical and electronic point of view) phosphorus atoms. These features inevitably influence upon chemical activity of P_4O_{10}; its interactions with nucleophilic reagents consist of several successive steps that differ considerably though formally they have in common the breakdown of P–O–P bonds.

Due to the high symmetry of the tricyclic structure and electron shielding of electrophilic phosphorus atoms by 10 oxygen atoms the first act of interaction of phosphoric anhydride with nucleophilic reagents is most likely to be the rate-limiting one. Initiation of such interaction needs either a rather high energy reactant or loosening of the symmetrical molecular electronic shell. The former could be achieved with the help of mobile hydrogen atoms. After the first few steps the whole process is likely to become autocatalytic which can be explained by two reasons. The first is the formation of the acidic P-O-H function which acts as an intramolecular catalyst for the disintegration of the tricyclic molecular structure of phosphoric anhydride. The other reason was based on molecular modelling of phosphoric anhydride which indicated that the product of the first step of a simple nucleophile addition produced a lower energy for the latter structure (9.66 kcal) which determined the rigidness of one cycle of phosphoric anhydride. So this first step of the reaction breaks down one of three cycles and initiates decomposition of the other two cycles due to lack of symmetry in the two cyclic residue of phosphoric anhydride and polarity of P–O–P bonds. Eventually the linear or branched tetraphosphoric structures contain three substituents probably at different phosphorus atoms and further reaction with nucleophilic regents is accompanied by break down of the 'ordinary' P–O–P bonds. The latter process is usually achieved by mild hydrolysis that leads to the formation of acidic phosphates.

CHAPTER 4

Phosphoric Anhydride in the Evolution of Life on the Primitive Earth

Phosphorus is considered to be the limiting nutrient for life [183], that is why the natural synthesis of organic hydroxyphosphates on the Primitive Earth was a prelude to the advent of life. Thus availability of reactive primodial phosphorylating reagents was of paramount importance [183]. In this respect scientists and evolutionists were actively speculating on the subject without the evidence for working out a reliable theory [183]. The main difficulty arose from the fact that practically all natural sources of phosphorus are sparingly soluble in water [581–583]. In recent decades an idea of prebiotic phosphorylation of organic compounds by polyphosphates was developed and intensively considered [130, 183, 584–586]. The most likely candidate for active phosphorylation on the Primitive Earth is tetrametaphosphate derived from phosphoric anhydride [183] (Chapter 3, Section 1.1) (Scheme 251).

$$(251)$$

Two principal ways of P_4O_{10} formation on the Primitive Earth [130, 183] were proposed. According to Newman and Neuman [586a], and Griffith and co-workers [183] the P–O–P bonds could arise from apatite and silica at temperatures above 2300°C [35]. Moreover apatite could act as a catalyst, scavenger and protector of the new P–O–P bonds.

Another approach to the generation phosphoric anhydride under the conditions of the Primitive Earth was a reaction of phosphorus-containing rocks with silica and carbon with the formation of phosphorus and further oxidation of phosphorus by carbon dioxide at high temperature [130] (Scheme 252).

$$2P + 5CO_2 \longrightarrow 0.5P_4O_{10} + 5CO \tag{252}$$

Thermodynamically this reaction is shifted to the right and its equilibrium constant was calculated from the standard free energy changes [130] at normal pressure and 1000°C (Scheme 253).

$$K = \frac{P_{CO}^5 \cdot P_{P_4O_{10}}^{1/2}}{P_{CO_2}^5 \cdot P_P^2} = 7.6 \cdot 10^5 \tag{253}$$

Could phosphoric anhydride exist under such conditions? Robie and co-authors [587] determined the free energy change for the hydration of phosphoric anhydride at various temperatures (Scheme 254. Figure 4.1).

$$1/2P_4O_{10} + 3H_2O \longrightarrow 2H_3PO_4 \tag{254}$$

According to Figure 4.1 one can see that the free energy changes of this reaction vary in direct proportion to temperature and P_4O_{10} should exist safely at the temperatures above 900 K and normal pressure even in a humid atmosphere.

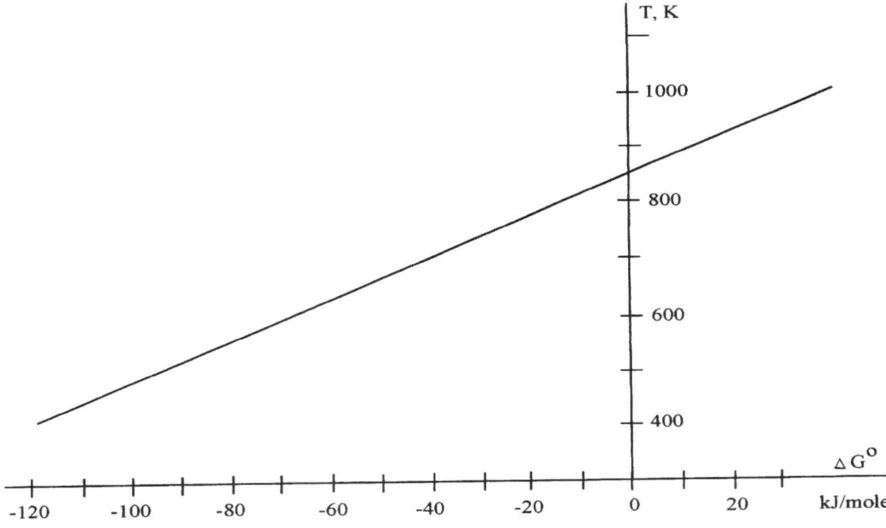

Figure 4.1 Free energy change of phosphoric anhydride at various temperatures and normal pressure

Another important question concerns the source of very high energy needed for the processes mentioned above. Griffith and co-workers [183] suggested lightning which struck outcrops of apatite, whereas Yamagata and co-authors [585] and Tageeva [588] suggested volcanoes based on the fact that phosphates were ejected due to the very high volcanic activity on the Primitive Earth.

One of the most significant properties of tetrametaphosphate from the point of the evolution theory is the solubility of its calcium and magnesium salts. (The authors of those theories do not mention isotetrametaphosphate as another product of phosphoric anhydride hydration.) After acting as a phosphorylating agent presumably in the primitive seas tetrametaphosphate could be recycled on the Primitive Earth [130] by precipitation to the seabeds in an insoluble form and acting further as a source of phosphoric anhydride formation when moved to the high energy zones by active crust alteration. Such processes are estimated to take place but to a less extent on Earth at present.

The validity of these theories was supported by Japanese scientists [130] who studied the phosphorylation of adenosine by a mixture of phosphoric anhydride with water. The dominating products of the reaction were 2'- and 3'-adenosinemonophosphates (AMP) as presented in Scheme 255. At low pH (i.e. values of less than 9.0) significant quantities of cyclic (2'–3')-AMP were formed. The formation of 2'–3'-adenosinemonophosphates via cyclic (2'–3')-AMP was confirmed by the authors [130].

$$\tag{255}$$

CHAPTER 5

Phosphoric Anhydride and its Derivatives in Biological Chemistry

Biochemists have used the benefits of phosphoric anhydride as a very efficient intra- and intermolecular dehydrating agent, a mild phosphorylating reagent and a catalyst for many decades. A number of examples are mentioned in Chapters 3 and 4. The application of phosphoric anhydride to bioorganic chemistry is derived from its general chemical reactivity as expressed in Chapter 3. This chapter presents some typical examples of bioorganic reactions involving P_4O_{10} or its derivatives. These examples include predominantly the reactions of direct phosphorylation, condensation and elimination.

From the biological point of view mononucleotides are the most important types of phosphoric acid monoesters [3]. One of the earliest examples concerning the phosphorylation of uridines by P_4O_{10} was presented by Hall and Khorana in 1955 [589]. The reaction of pyrimidine nucleosides with phosphoric anhydride and phosphoric acid (85%) without a solvent followed by treatment with an alkali resulted in the formation of a mixture containing uridine-5' phosphate and uridine -2'(3'),5 diphosphate (Scheme 256).

The nature and the ratio of products were time dependent but in all cases the former products were dominating. Presumably the diphosphate was formed via monophosphate because the generation of diphosphate only started after one hour when the level of monophosphate was beyond 60%. Uridine-2'(3'),5 diphosphate was synthesized also by phosphorylation of nonprotected ribonucleoside with a similar phosphorylating mixture [589].

Schramm and co-workers were the first to be interested in Langheld ester (polyphosphoric ester which is prepared from phosphoric anhydride and diethyl ether, Chapter 3, Section 1.3). They used it as a condensing agent for the synthesis of nucleosides, polysaccharides, nucleic acids [590] and

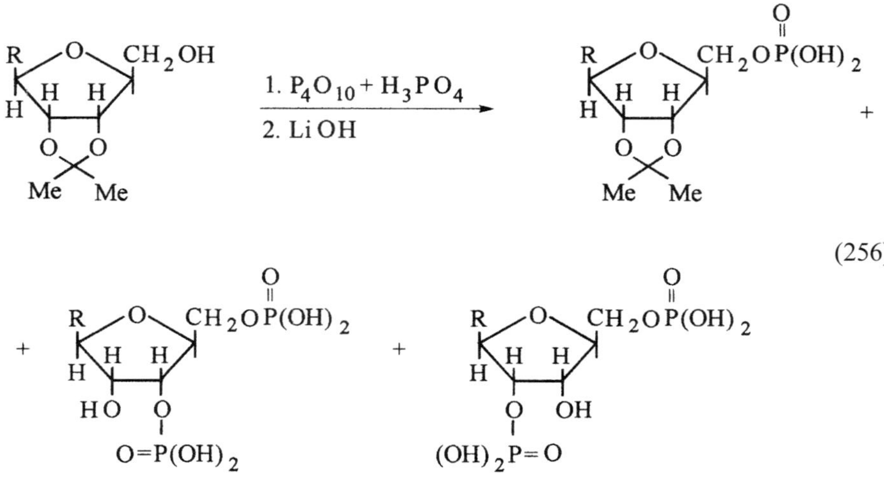

$$\text{(256)}$$

R = adenine, uracil, cytosine

polypeptides [591]. For example polyphosphoric ester promoted the condensation of ribose with adenine [590] as presented in Scheme 257.

$$P_4O_{10} + EtOEt \longrightarrow \text{Polyphosphoric ester (PPE)} \qquad (\text{Section } 3.1.3.)$$

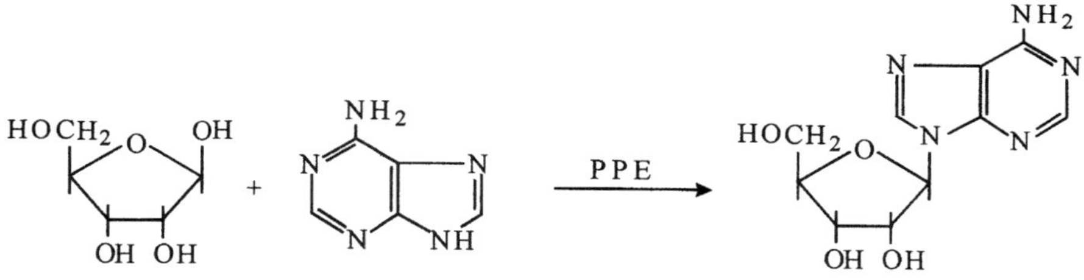

$$\text{(257)}$$

The optimum ratio of ribose and polyphosphoric ester was found to be 1:2. The most reactive hydroxyl groups in glucose, fructose and methylglucoside were at the positions C1 and ω [146]. On the bases of experimental data it was [146] concluded that the interaction between sugars and Langheld ester proceeded via phosphorylation of the former atom.

A few years later Onodera, Hirano and Fukumi [592] synthesised O-glycosides, glycosylamines and purine nucleosides using reactions initiated by polyphosphoric acid or ethyl polyphosphate. The fully acetylated sugars (1.0 mole) were mixed with an aromatic alcohol (phenol, p-cresol) (1.2 moles) or amine (aniline, p-toluidine, m-nitroaniline, theophylline and 2-naphthylamine) (1.2 moles) and fused with a catalytic amount of polyphosphoric acid or ethyl polyphosphoric ester, that were claimed to be the new catalysts, at 100–160°C for 30–60 min. The yield of glycosides was dependent to some extent upon the ratio of the reactants. For example the reaction did not take place without the catalyst but in the range of its concentration from 1–50% the yields of the products varied irregularly between 61 and 82% [592]. The optimum amount of ethyl polyphosphoric acid in the process between 1,2,5-tri-O-acetyl-β-D-glucofuranurono-6,3-lactone and phenol to give the maximum yield (82%) of phenyl 2,5-di-O-acetyl-β-D-glucofuranurono-6,3-lactone was 3.0%. A similar effect on the reaction was established for polyphosphoric acid as well. The nature of the catalysts also had a considerable impact on the results of the synthesis [592]. The new catalysts ethyl polyphosphate and polyphosphoric acid were tested against the traditionally used p-toluenesulphonic acid and zinc chloride in the reactions of 1,2,5-tri-O-acetyl-β-D-glucofuranurono-6,3-lactone with phenol and 1,2,3,4,6-penta-O-acetyl-β-D-glucopyranose with theophylline. In the former process polyphosphoric acid and ethyl polyphosphate gave a threefold larger yield than p-toluenesulphonic acid. In general, the polyphosphates gave yields that were 6–13% higher than that of p-toluenesulphonic acid and *c.* 3.5 times larger than zinc chloride. The authors [592] emphasized the significant role of polyphosphoric acid and ethyl polyphosphate catalysis for the anomeric configuration of the produced glycosides and worked out the relationship between the acidity of the aromatic reagents and anomeric configuration. The pK_a value greater than 5.07 of p-cresol, phenol, theophylline and p-toluidine induced the formation of β-D-glucosides, whereas aniline, 2-naphthylamine and m-nitroaniline characterized by the pK_a value 4.58 and less gave α-D-glucosides or in other words these catalysts could produce both types of anomers depending upon the acidity of phenols or amines. All attempts to employ polyphosphoric acid and ethyl polyphosphate for the synthesis of pyrimidine nucleosides from 1,2,3,4,6-penta-O-acetyl-β-D-glucopyranose with uracil, 2,4-dimethoxypyrimidine and 4-methoxy(ethoxy)-2-(1H)-pyrimidinone gave no products of condensation [592].

The more recent research data [153, 154] were in considerable disagreement with those presented by Schramm and co-workers [146, 590, 591]. Synthesis of uridine 2′(3′)-phosphate polymer and copolymers from adenosine, cytidine, uridine 2′(3′)-phosphate and adenylic, uridilic, cytidilic acids and some other compounds showed that irregular polymers and non-natural types of linkages were generated in the condensation reactions initiated by ethyl polyphosphoric ester, leading to polynucleotides.

The 2′(3′),5-diphosphate of the nucleoside X and N^3-alkylated uridine derivatives were synthesized in the reaction of the corresponding nucleosides with pyrophosphoryl tetrachloride [593]. The latter reagent was prepared from phosphoric anhydride and phosphorus pentachloride in carbon tetrachloride at 105–110°C within 8 h (Scheme 258).

$$P_4O_{10} + PCl_5 \longrightarrow Cl_2P(O)-O-P(O)Cl_2 \quad (50\%) \tag{258}$$

In order to achieve a high yield the phosphorylation of nucleosides required 15-fold excess of the phosphorylating agent. Tetrachloropyrophosphate was used for the phosphorylation of some other nucleosides as well [594].

Phosphoric anhydride has also been utilised to initiate a condensation reaction leading to the formation of tripeptides and oligopeptides in diethyl phosphite media [591]. The procedure was developed by a study of the reaction of aniline with phosphoric anhydride in diethyl phosphite. It was proposed that the first step of the reaction was the formation of ethyl polyphosphate from P_4O_{10} and $(EtO)_2HP=O$ followed by phosphorylation of aniline (or corresponding peptides) and condensation of the phosphorylated product with a substituted amino acid (Scheme 259), using the phosphoryl group as a leaving group.

$$RNHCR^1COOH + EtOP(O)(OH)NHCR^2COOR^3$$
$$\longrightarrow RNHCR^1C(O)NHCR^2COOR^3 + EtOP(O)(OH)_2 \tag{259}$$

As early as 1909 Fischer and Delbruck [595] were the first to study the influence of phosphoric anhydride upon the synthesis of oligosaccharides from substituted monosaccharides applying the basic dehydrating activity of P_4O_{10}. The same approach was used by later research groups. For example Micheel and co-authors [596] showed that phosphoric anhydride could successfully be used in place of thionyl chloride in the process whose catalytic activity was due to the formation of hydrochloric acid.

Another efficient method successfully used for saccharides and oligosaccharides polymerization was based on the use of P_4O_{10} in acidic [598] or neutral [597, 599–603] solutions of chloroform–dimethylsulphoxide. A typical reaction lasted for 2–15 days at 30–37°C to give yields of glycans from 6 up to 34% [597]. Heating at higher temperatures (60–65°C) led to oxidation [600]. The main concern was the structure of polysaccharides produced under the action of phosphoric anhydride in dimethylsulphoxide. Micheel and Bockmann [596] presented the linear structure for 2,3,4-trimethylpolyglucosan with entirely 1,4-linkages and having an average molecular weight of 5600. The main concern of Husemann and Muller [603], Mizuro [604] and some other authors also concerned the structure of polysaccharides, which provoked many speculations. The routine experimental data (degradation of α-amilase, optical rotation in TEAH and viscosity) illustrated the formation of entirely β-1,4-glucosidic linkages in the polymers, the degree of polymerization in which was up to 640 [603]. Hirano and co-authors [600] concluded that a branched structure of a polysaccharide was built up under the action of phosphoric anhydride on monomeric carbohydrates pointing out the evidence for α-1,4- and α-1,6-glucosidic linkages in the main chain of synthetic glucan. A very important result of the experiments was the identification in the product of polysaccharides containing a significant amount of phosphorus (1.3–20.9%) which could not be removed by mild methanolysis [600, 603]. The presence of the phosphoryl group in the polymeric product was supported also by spectroscopic methods. The role of phosphoric anhydride in the saccharide polymerization process is said to be not entirely dehydrating but catalytic as well [600]. Presumably phosphoric anhydride phosphorylated monomeric saccharides which was followed by intermolecular condensation with the phosphorus moiety acting as a leaving group.

The same combination of phosphoric anhydride with dimethylsulphoxide was found to be a selective oxidizing agent for carbohydrates [605]. Unlike fully substituted carbohydrates such as 7-(tetra-O-acetyl-β-D-glucopyranosyl) theophyline, phenyl-tetra-O-acetyl-α-D-glucopyranoside, tri-O-acetyl-1,6-anhydro-β-D-glucopyranose, 5-O-acetyl-1,2O-isopropylidene-α-D-glucofura-nurono-6,3-lactone, methyl 4,6-O-benzylidene-α-D-glucopyranoside 2,3-dinitrate and some others, two glucosylamines — N-p-tolyl-2,3,4-tri-O-benzoyl-L-arabinosylamine and N-p-tolyl-2,3,4,6-tetra-O-acetyl-D-glucopyra-nosylamine were oxidised by P$_4$O$_{10}$–Me$_2$SO. According to these results the following substituents — acetoxy, sulphonyloxy, benzoyloxy, isopropylidene, benzylidene, ethylidene, methoxyl, nitrate and acetylamino group as well as glycosidic bonds possessing nucleosidic, phenolic, alkyl and thioacetyl fragments were determined to be suitable protective groups against oxidation, hence the carbohydrates with the above protecting groups could be introduced safely in the synthesis of dicarbonyl sugars [605]. The experimental conditions were varied across a wide range of temperatures, ratio of components in the oxidizing mixture and different solvents. The process of ketosugar formation shown in Scheme 260 was monitored by thin layer chromatography and NMR spectrometry.

$$\text{(260)}$$

Heating the reagents mixed with the ratio 1 mole of sugar per 1–1.5 moles of phosphoric anhydride and 3–4 moles of dimethylsulphoxide in dimethylformamide at 65–70°C gave the maximum yield (85–92%) of the product [605]. If dimethylsulphoxide was used as a solvent at an elevated temperature the yield was reduced due to decomposition. A large excess of phosphoric anhydride also decreased the yield of the product. In addition, catalytic quantities of P_4O_{10} were ineffective in promoting the oxidation process [605]. The most efficient solvents were found to be dimethylsulphoxide and dimethylformamide. For example, the partial substitution of these solvents by either benzene or chloroform resulted in a pronounced reduction in the yields of the keto derivative. Attempts to catalyse the reaction with diantimony trioxide or pentoxide, diarsenic trioxide or pentoxide and p-toluenesulfonic acid were unsuccessful.

A group of British researchers [606] performed the efficient oxidation of 1,2:5,6-di-O-isopropylidene-α-D-glucofuranose into 1,2:5,6-di-O-isopropylidene-α-D-ribo-hexafuranos-3-ulose, 1,2:3,4-di-O-isopropylidene-L-rhamnitol into 1-deoxy-3,4:5,6-di-O-isopropylidene-L-fructose and methyl 4,6-O-benzylidene-2-O-toluene-p-sulphonyl-α-D-glucopyranose into methyl 4,6-O-benzylidene-2-toluene-p-sulphonyl-α-D-ribo-hexopyranosid-3-ulose by heating the initial carbohydrates with phosphoric anhydride for 48 h at 25–50°C.

The presence of the carbonyl group in the amino carbohydrates opened new synthetic routes to isomeric, branched-chain and diamino sugars which makes the above reactions of considerable importance.

Scheme 261 shows the preparation of a ketose by the oxidation of the corresponding carbohydrate using phosphoric anhydride–dimethylsulphoxide in dimethylformamide with the yield 71% [606].

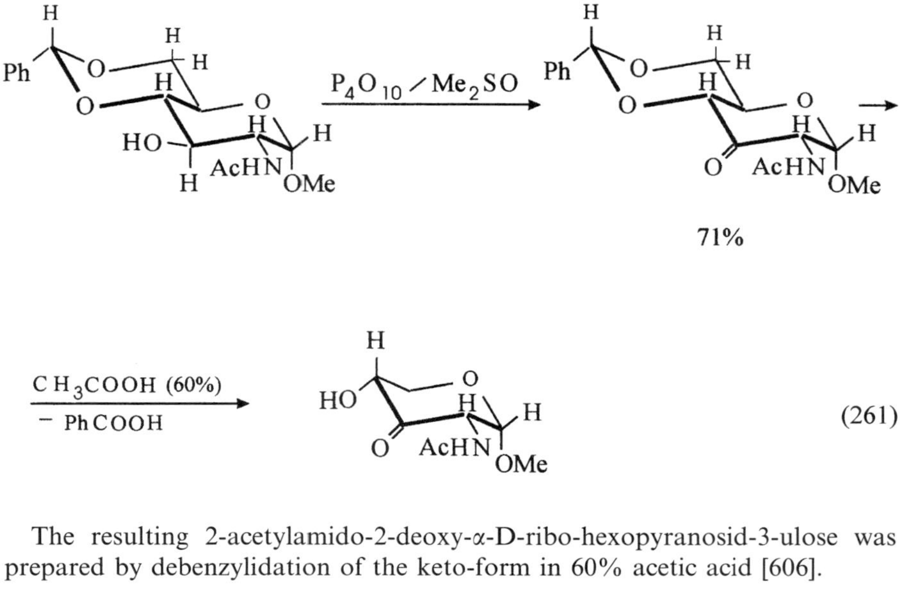

$$(261)$$

The resulting 2-acetylamido-2-deoxy-α-D-ribo-hexopyranosid-3-ulose was prepared by debenzylidation of the keto-form in 60% acetic acid [606].

A 65% yield of 1,2:5,6-di-O-isopropylidene-α-D-ribo-hexofuranos-3-ulose (Scheme 262) was obtained using two molar equivalents of phosphoric anhydride in dimethylsulphoxide at 60–65°C for 1.5–2 h [606].

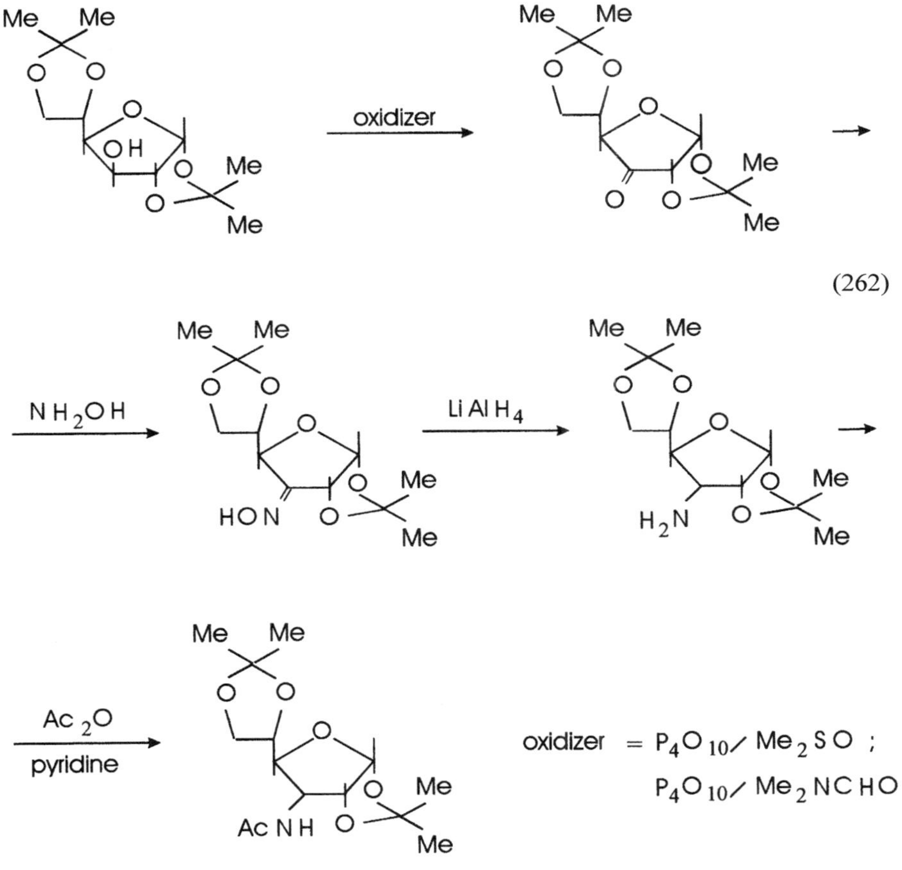

$$(262)$$

oxidizer $= P_4O_{10} / Me_2SO$;

P_4O_{10} / Me_2NCHO

Onodera and co-authors [605] found this discovery of particular importance due to the fact that 1,2:5,6-di-O-isopropylidene-α-D-glucofuranose had exhibited no reactivity towards dimethylsulphoxide–DCC reagent [607, 608] nor towards many other traditionally oxidising reagents with the exception of ruthinium tetroxide [609].

The isolated ketose was reacted with hydroxylamine, thiosemicarbazide or semicarbazide to produce the corresponding crystalline oxime (Scheme 262) which was subsequently hydrogenated with lithium aluminium hydride to give the amino derivative. Subsequent acylation by acetic anhydride in pyridine gave 3-acetamido-3-deoxy-1,2:5,6-di-O-isopropylidene-α-D-allofuranose [609] (Scheme 262).

The most remarkable aspect of the carbohydrate oxidation by phosphoric anhydride–dimethylsulphoxide mixture is the insignificant steric demand on the oxidants. For example, P_4O_{10}:Me_2SO also oxidized hindered secondary alcohols [609] as shown in Scheme 263.

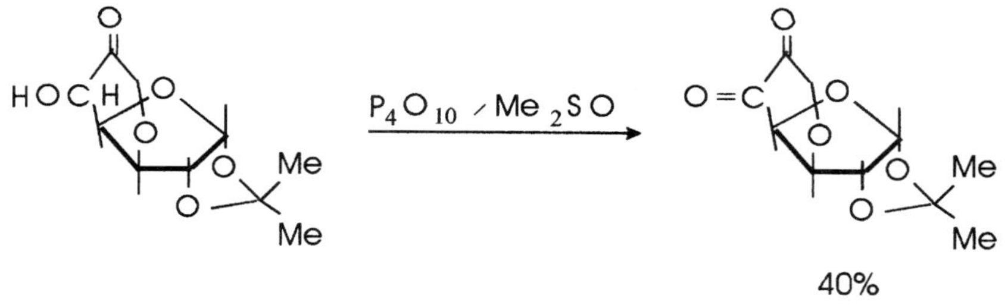

$$P_4O_{10} \diagup Me_2SO$$

40%

(263)

The fairly efficient oxidation of sterically hindered epimeric pairs of sugars revealed that the oxidizing mixture was equally effective towards both epimeric derivatives [609]. The orientation of the hydroxyl group had a little effect if any on the rate of oxidation by phosphoric anhydride–dimethylsulphoxide which was quite unusual for the oxidation of sugars.

2-Cyanoethyl phosphate has been used in biochemical reactions as an efficient phosphorylating agent. Usually it had been synthesized in low yield by the reaction of 2-cyanoethanol and phosphorus oxychloride followed by hydration in the presence of pyridine and barium carbonate. Terner [610] developed an efficient synthetic approach to 2-cyanoethyl phosphate from 2-cyanoethanol, phosphoric anhydride and phosphoric acid in the presence of acetic anhydride (Scheme 264). The method involved the use of excess dicyclodiimide in pyridine to couple the phosphoric acid mono cyanoethyl ester to various nucleotides. The cyanoethyl group in the intermediate diester was preferentially cleaved by very mild alkaline hydrolysis to give the desired 3-phosphate ester. In this way, for example, thymidine-3′-phosphate was prepared by phosphorylation of 5′-O-tritylthymidine (Scheme 265).

$$HOCH_2CH_2CN + P_4O_{10} + H_3PO_4 \xrightarrow{Ac_2O} (HO)_2 \overset{\overset{\displaystyle O}{\|}}{P} - OCH_2CH_2CN$$

(264)

$$\text{Ph}_3\text{COCH}_2\text{O-(sugar base, }Me\text{, }OH\text{)} \quad + \quad (\text{HO})_2\overset{\text{O}}{\overset{\|}{\text{P}}}-\text{OCH}_2\text{CH}_2\text{C N} \quad \longrightarrow$$

(265)

$$\longrightarrow \quad \text{Ph}_3\text{COCH}_2\text{O-(base)}-\text{O}-\overset{\text{O}}{\underset{\|}{\underset{\text{O}}{\text{P}}}}-\text{OCH}_2\text{CH}_2\text{C N} \quad \xrightarrow[\text{2. OH}^-]{\text{1. H}^+} \quad \text{Ph}_3\text{COCH}_2\text{O-(base)}-\text{O}-\text{P(OH)}_2\!=\!\text{O}$$

Phosphoric anhydride has played a key role in the synthesis of the group B vitamins as well as their phosphorylated and nonphosphorylated derivatives and analogues.

Phosphoric anhydride or its mixture with phosphoric acid provides an experimentally convenient and efficient route to the synthesis of pyridoxal-5′-phosphate. This phosphate has been claimed to be a substitute for vitamin B$_6$ due to its participation as the coenzyme in numerous enzymatic processes with amino acids including transamination, decarboxylation, desmolysis, racemization and some others [611]. For this reason the reaction of pyridoxal phosphorylation has been studied for decades. The earliest investigations involved the common phosphorylating agents such as phosphorus oxychloride [612, 613] and metaphosphoric acid [614–616] but without great success because of either very low yields of pyridoxal-5′-phosphate or its isolation in an impure form.

The first attempt to phosphorylate pyridoxal employing phosphoric anhydride was described by Baddiley and Mathias [617]. The isopropelidene derivative of pyridoxine was reacted with the phosphoric anhydride–phosphoric acid mixture and the phosphorylated intermediate was hydrolysed to pyridoxine phosphate which eventually was oxidized by manganese dioxide at elevated temperature to give pyridoxal-5′-phosphate with a very low yield (Scheme 266).

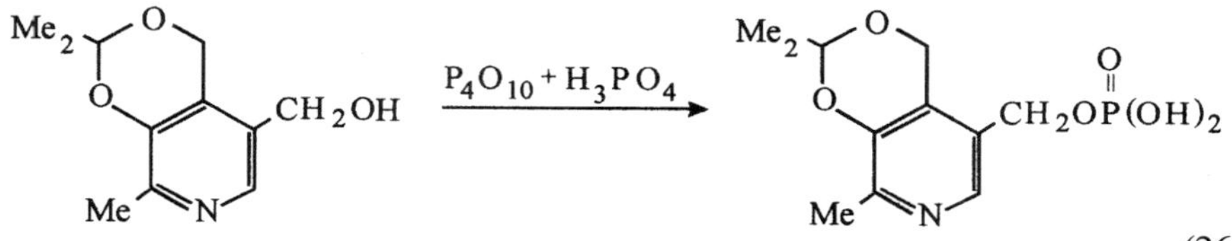

(266)

The disadvantages of the above method, i.e. low yield and insufficient purity of the product, were solved by Peterson and Sober [618] who modified the synthesis by using pyridoxamine rather than pyridoxine as the substrate for the phosphorylation by the mixture P_4O_{10}/H_3PO_4 (Scheme 267).

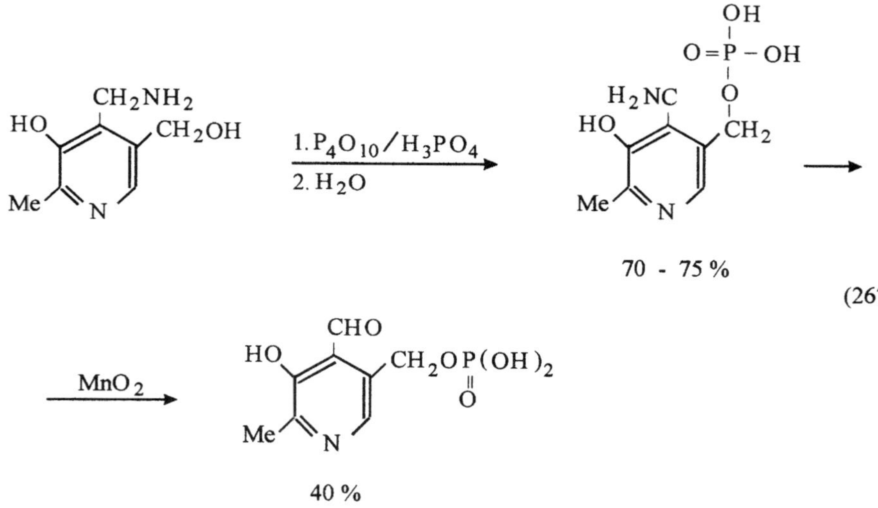

(267)

If pyridoxamine was reacted with neat phosphoric anhydride in dry media rather than with the mixture P_4O_{10}/H_3PO_4 it would have probably phosphorylated the more nucleophilic amino group, and not the hydroxymethylene group at position 5 of the cycle (Chapter 3, Section 3). This example demonstrates the pronounced difference between the two phosphorylating agents.

From the synthetic point of view the most complicated step of the process (Scheme 267) was the oxidation of pyridoxamine-5′-phosphate by manganese dioxide. It was discovered [618] that only a freshly made very active form of the oxidizing reagent in dilute aqueous solution (1–10 mg/ml) at pH about 5.0 for 30–60 min gave pyridoxal-5′-phosphate in good yield. Three–fivefold excess of manganese dioxide gave the product in a pure form with only traces of impurities.

A still more advanced approach to pyridoxal-5'-phosphate was patented by Schorre [619]. The principal invention involved protection of the aldehyde group, prior to phosphorylation by P$_4$O$_{10}$/H$_3$PO$_4$, i.e. phosphorylation via the oxazolidine derivative of pyridoxal (Scheme 268). The protected compound was synthesized in two ways: in the reactions of substituted ethanolamines with either pyridoxal or hydroxysulphonic acid as presented by Scheme (268).

R = H, Alk, Alk - cyclo, Ar, Alkylaryl ;

$$\text{P} = \overset{\displaystyle O}{\underset{\displaystyle OH}{\overset{\|}{P}}} - O - \overset{\displaystyle O}{\underset{\displaystyle OH}{\overset{\|}{P}}} - O - \overset{\displaystyle O}{\underset{\displaystyle OH}{\overset{\|}{P}}} - O - \overset{\displaystyle O}{\overset{\|}{P}}(OH)_2$$

$$(268)$$

The first step of this process involved benzene isotrope to remove water at 65–70°C. Phosphorylation of pyridoxal oxazolidine by P$_4$O$_{10}$/H$_3$PO$_4$ involved heating at 60°C for 2 hrs. The formation of tetraphosphate under the conditions as suggested by the author [619] seems over simplified because a mixture of phosphoric anhydride with phosphoric acid consists of a mixture of poly-, tetra-, tri-, pyro- and orthophosphoric acids which could result in formation of a number of phosphorylated intermediates. Subsequent hydration of the phosphorylated pyridoxal oxazolidine was achieved by hydrochloric acid (0.1 N). No yields of the individual steps were given [619], nor was there an explanation of the advantages of this method over earlier developed methods.

Biotin is the water-soluble vitamin H or coenzyme R. It is of particular importance as a coenzyme in reactions of carboxylation related to various biochemical processes including gluconeogenesis and biosynthesis of fatty acids [620]. There are many natural sources of biotin such as liver, kidney, pancreas, milk, egg yolk and others but the deficiency of this coenzyme causes many severe symptoms. Thus the development of an efficient synthetic approach to biotin is a challenge for chemists that has led to the development of a number of state-of-the-art methods. The crucial step of one of the methods involved elimination of thiomethanol from substituted methylthiotetrahydrothiophene by a mixture of phosphoric anhydride and sulphuric acid [621] (Scheme 269).

$$\xrightarrow[\substack{-\,MeSH \\ (0^{\circ}C,\ C_5H_5N)}]{P_4O_{10}/H_2SO_4} \qquad 73\%$$

$$\xrightarrow{4\ \text{Steps}}$$

Biotin

(269)

The authors [621] emphasized that the application of sulphuric acid without phosphoric anhydride led only to the substitution of the thiomethyl group by a hydroxyl group. The last four steps involved reduction of the keto group, hydrolysis, cyclization of urethane and dehydration eventually gave biotin with a good overall yield [620, 621].

Thiamine hydrochloride (vitamin B_1) reacted with the phosphoric anhydride–phosphoric acid mixture at 100–120°C to give a mixture of the corresponding monophosphate (37.2%) and diphosphate (33.7%) [622].

Phosphorylated azoles that can be produced in the corresponding reaction with P_4O_{10} have been used as specific phosphorylating agents particularly in biochemical synthesis. For example Kraszewski and Stawinski [623] used phosphoryl tris-1,2,4-triazole for introducing the phosphoryl group in 5′-dimethoxytritylthymidine (Scheme 270).

$$\text{DMT-T-OH} + \text{O=P}\left(\text{N}\underset{\text{N}}{\overset{\text{N}}{=}}\right)_3 \longrightarrow \text{DMT-T-O}-\overset{\text{O}}{\overset{\|}{\text{P}}}\left(\text{N}\underset{\text{N}}{\overset{\text{N}}{=}}\right)_2 \longrightarrow$$

(270)

$$\xrightarrow[\text{2. R}^2\text{OH}]{\text{1. R}^1\text{OH}} \quad \text{DMT-T-O}-\overset{\text{O}}{\underset{\overset{|}{\text{OR}^2}}{\overset{\|}{\text{P}}}}-\text{OR}^1$$

Phosphoric anhydride was used successfully by Knight and Pattenden [624, 625] in the final step of the synthesis of pulvinic acid pigments of lichen and fungi. The total synthesis with unspecified yield is illustrated by Scheme 271 for the permethylated derivative of gomphidic acid which is the pigment of *Gomphidius glutinosur* [621].

$$(271)$$

The process of dehydration of the tertiary alcohol was accomplished quite simply, although rather slowly, by heating in benzene. The same method was employed for the synthesis of O-methyl pinastric acid [621].

Thiophosphoryl tris-imidazole is a phosphorylating agent for 5'-acylthymidine [348]. Phosphorylated azoles were efficiently used for oligonucleotide synthesis, phosphorylation of peptides and other biologically active compounds [349]. In our experience it is likely that monotriazolyl phosphoric acids would also act as a phosphorylating agent which would produce directly (without a consequent step of hydrolysis) the biologically active compounds containing the dihydroxyphosphoryl substituent.

Three days' treatment of proteins with phosphoric anhydride and phosphoric acid at room temperature resulted in the phosphorylation of proteins — attack occurring at the aliphatic hydroxyl groups of serine, threonine, hydroxyproline and to a lesser extent tyrosine [626]. The authors found many common features between the phosphorylation of proteins by phosphoric anhydride and sulfuration by concentrated sulphuric acid. The possible phosphorylation of the amino groups in proteins was rejected. The highest degree of phosphorylation was achieved for gelatin. Phosphorylated proteins exhibited considerable stability and retained the phosphate group across the range of pH between 2.2 and 11.5 [626] (Scheme 272).

$$\textit{\textbf{PROTEIN}}-\text{OH} \quad \xrightarrow[\text{r. t. , 3 days}]{P_4O_{10}\,/\,H_3PO_4} \quad \textit{\textbf{PROTEIN}}-\text{O}-\textcircled{P}$$

$$\textcircled{P} \; = \; \text{Metaphosphate , Orthophosphate} \tag{272}$$

The selective action of P_4O_{10}/H_3PO_4 on proteins is usually quite different from the processes involving other phosphorylating agents including phosphorus oxychloride that lead to phosphorylation of both hydroxyl and amino groups [626].

N-Formyl-d,l-tryptophan prepared by acylation of triptophan with acetic anhydride in formic acid was subjected to dehydration–cyclization under the action of polyphosphoric acid in the presence of phosphorus oxychloride at 125°C (80 min). The process was accompanied by vigorous evolving of CO_2 and HCl. On a routine work-up the alkaloid *harmon* was isolated [627] (Scheme 273), which is quite difficult to synthesize by other methods.

In a similar manner polyphosphoric acid caused the dehydration–cyclization of N-acetyl-d,l-phenylalanine as presented by Scheme 274.

Comparison of the processes shown in Schemes 273 and 274 and the reactions of polyphosphoric acid presented earlier (Chapter 3, Section 8.1) illustrates well the influence of the amide structure upon the reaction pathways.

(273)

35 %

(274)

92 %

However, the method is not always applicable as was shown by Galat [628]. Attempts to cyclodehydrate N-(3′,4′-dimethoxyphenylacetyl)-3,4-dimethoxy-phenyl alanine by polyphosphoric acid aimed at the preparation of *papaverine* (Scheme 275) were unsuccessful [629].

(275)

Intensively used as a highly efficient natural muscle relaxant and analgesic, the chemistry of the nonaddictive narcotic *papaverine* has been a source of many challenges for chemists. Early attempts at the synthesis of *papaverine* were found to be particularly difficult and numerous attempts to prepare it resulted in failure; an example of which is given above. Eventually Pictet and Gams discovered a new type of reaction (the Pictet–Gams reaction) which led to the synthesis of *papaverine*. The last two steps were crucial, involving the transformation of ketoamine into papaverine. This was achieved by hydrogenation of the keto group and one-pot dehydration of alcohol and dehydration–cyclization of the amido group in accordance with Scheme (276).

$$(276)$$

Papaverine

Despite the numerous advantages that P$_4$O$_{10}$ presents to the synthetic approach to phosphorylated biologically active systems many chemists prefer to apply more complicated methods for phosphosaccharides, mononucleotides and phospholipids via chlorophosphates such as phosphorylchloride [628a, 629], dialkyl and diaryl chlorophosphates [630], dialkylamido and diarylamido chlorophosphates [631]. For example, one of the efficient 'Phosphoryl Chloride' methods used by Michelson and Todd [632] involved dibenzyl-chlorophosphate (Scheme 277).

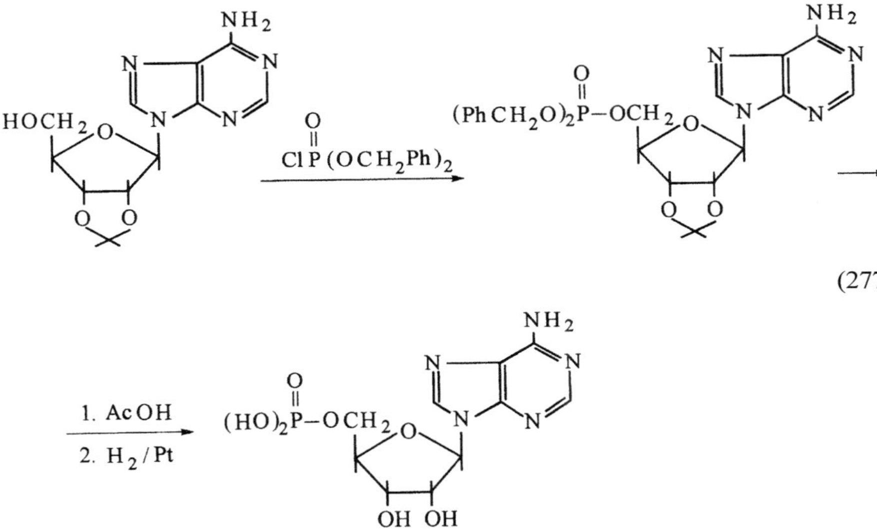

$$\text{(277)}$$

The most attractive 'nonphosphoric anhydride' approach to a limited number of phosphorylated compounds, including biologically active ones, was proposed by Polish chemists [631a]. Trialkyl and triaryl phosphites were phosphorylated by orthophosphoric acid in nitromethane with total yields of 80–90% [632] (Scheme 278).

$$\text{P(OR)}_3 + \text{H}_3\text{PO}_4 \longrightarrow \text{ROP(O)(OH)}_2 + \text{(RO)}_2\text{P(O)H} \qquad (278)$$

For obvious reasons this impressive method for acidic phosphates is inapplicable in some cases.

Zavlin and Sokolovski [633] studied the reaction of aminoalcohols that contained two competing nucleophilic groups with phosphoryl chlorides including phosphorus oxychloride (Scheme 279). The process led to the formation of phosphates with good biological activity.

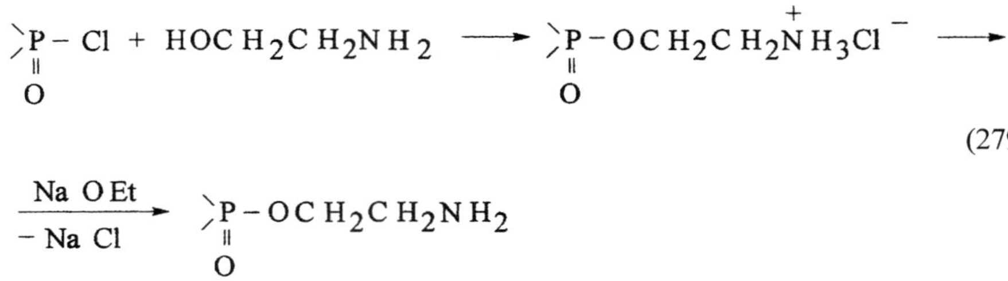

$$\underset{O}{\overset{}{>}}\!\!P\!-\!Cl + HOCH_2CH_2NH_2 \longrightarrow \underset{O}{\overset{}{>}}\!\!P\!-\!OCH_2CH_2\overset{+}{N}H_3Cl^- \longrightarrow$$

(279)

$$\xrightarrow[-\,Na\,Cl]{Na\,OEt}\ \underset{O}{\overset{}{>}}\!\!P\!-\!OCH_2CH_2NH_2$$

The results of this reaction are in good agreement with the experimental data described by Grun and Limpacher [90–93] and which are presented in Chapter 3, Section 1.2.

Several pyrimidinetriones are phosphorylated by phosphoric anhydride (Chapter 3, Section 3.4.2) to give products which have a pronounced soporific effect [634].

The total synthesis of ($\pm$)-equilenin methyl ether and ($\pm$)-isoequilenin methyl ether was accomplished [635] (Schemes 280, 281) by the application of a *phosphoric anhydride–phosphoric acid–phosphorus oxychloride* mixture for the key cyclization step (Scheme 281).

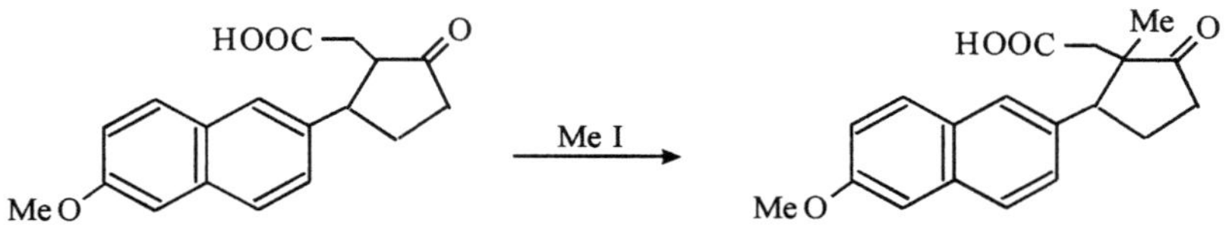

(280)

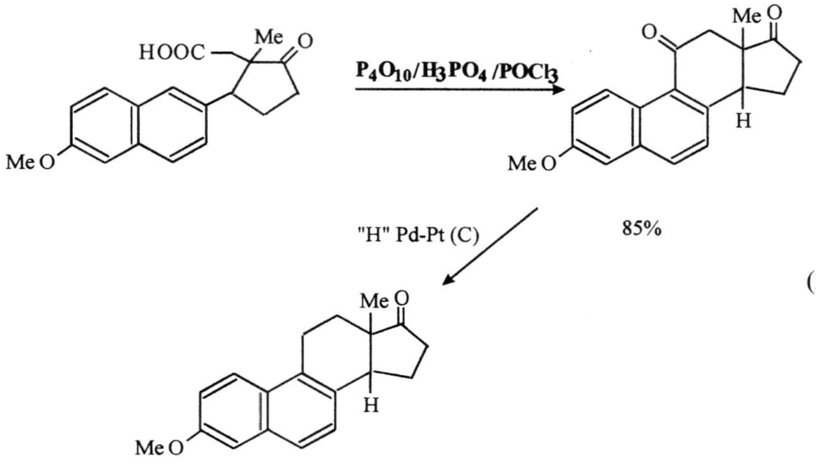

(281)

The cyclization was completed within 1.5 min at about 100°C with a much higher yield than had been achieved in earlier studies under the action of different reagents. The overall yield of equilenins was also very good.

CHAPTER 6

Applications of Phosphoric Anhydride

1 ENVIRONMENTAL APPLICATIONS

The organic phosphates, pyrophosphates, polyphosphates and most of their derivatives possess dozens of very useful properties that gives them a wide range of applications. There is also another extremely valuable property of phosphates, which is their low toxicity in combination with biodegradability.

Aaronson [636] gave a persuasive discussion supporting organic phosphate flame retardants as an alternative to many other types of traditionally used additives. The Dutch Ministry of the Environment intend to legislate for a reduction and possibly prohibit the application of flame retardants such as organobromine compounds and antimony oxide (that are still widely advertised) [637]. The problem not only concerned their acute and chronic toxicity but also their degradation products; i.e. ecotoxicity of the compounds in streams, soil and ground water [636].

The US Navy has banned the use of plastics containing organohalogen flame retardants after the Falklands war when electronics were damaged by corrosive gases formed during shipboard fires [636]. The use of some other types of organic compounds was also restricted predominantly for environmental reasons opening up to the market for organic phosphates.

1.1 IONITES WITH THE PHOSPHATE GROUP

The ion exchange and particularly cation exchange resins containing the phosphate group exhibit valuable properties such as high selectivity, thermal stability and mechanical strength. For example polyvinyl alcohol is modified for this purpose by partially esterifying it using phosphoric acid [638] (Scheme 282).

$$\{CH_2 - \underset{|}{CH}\}_n + mH_3PO_4 \xrightarrow[-H_2O]{} \{CH_2 - \underset{|}{CH}\}_{\overline{n-m}}\{CH_2 - \underset{|}{CH}\}_m$$
$$\qquad\qquad OH \qquad\qquad\qquad\qquad\qquad OH \qquad\qquad OPO_3H_2$$

$$(282)$$

Hayashi [639] prepared similar resins by the polymerization of diethyl vinyl phosphoric ester followed by alkali hydrolysis of the alkyl phosphoric groups.

Phosphoric anhydride possesses high potential in the preparation of phosphorylated ionites. For example the treatment of polyvinyl alcohols, cellulose [112, 113], starch [110] and other polymers with proton-donating properties (Chapter 3, Section 1.2, Chapter 6, Section 4) by phosphoric anhydride gives the corresponding polymers with the dihydroxyphosphoryl group possessing pronounced ion-exchange properties.

1.2 WATER CONDITIONING AND CONTROLLING

Several patents utilize phosphoric anhydride in the production of water softeners and conditioners. The products from the phosphoric anhydride reaction with ammonia (see Chapter 3, Section 3.1) was patented for this application as long ago as 1938 [172]. Water-soluble glasses were made by the combination of phosphoric anhydride and silica with a source of either potassium oxide [640] or sodium oxide [641]. Water treatment processes have also utilized water-soluble glass of the type $Na_2O \cdot Al_2O_3 \cdot P_2O_5$ [642]. Calcium polyphosphates such as $Na_2O \cdot CaO \cdot P_2O_5$ have been recommended as corrosion inhibitors in boilers [643]. Similar uses have been patented for water-soluble phosphates made from calcium metaphosphate and sodium or ammonium sulphate [644].

Mono-, di-, tri- and polyalcohols phosphorylated by phosphoric anhydride or their products of condensation with ethylene oxide or propylene oxide (Scheme 283) were patented [645] as active inhibitors of scale depositing from hard water. The phosphorylation was carried out at 95–150°C (sometimes 130–135°C) for 30 min. Sometimes heating for 3–5 h was necessary to complete the process. A complex of the product with boron trifluoride etherate was used as a catalyst.

The overall process for most organic reagents studied by Vogelsang [645] is shown in Scheme 283. The alcohols studied were glycerol, sorbitol, mannitol, erythritol, arabitol, xylitol, quercitol, inositol and mono-, di- or tri-pentaerythritol. Secondary hydroxyl groups were naturally less active towards phosphoric anhydride than the primary ones. Some lower alcohols such as propanol were added to the reaction mixture to achieve the desired viscosity. The lower phosphorylated alcohols were found to be not particularly active as scale inhibitors.

$$(CH_2O)_xH$$
$$(CH_2O)_yH$$
$$(CH_2O)_xH$$

$$\xrightarrow{\overset{O}{RCH - CH_2}}$$

$$(CH_2O)_x(CH_2\overset{R}{CHO})_nH$$
$$(CH_2O)_y(CH_2CRHO)_mH$$
$$(CH_2O)_x(CH_2\underset{R}{CHO})_pH \longrightarrow$$

(283)

$$\xrightarrow{P_4O_{10},\ Et_2O\cdot BF_3}$$

$$(CH_2O)_x(CH_2\overset{R}{CHO})_n-\text{(P)}$$
$$(CH_2O)_y(CH_2CRHO)_m-\text{(P)}$$
$$(CH_2O)_x(CH_2\underset{R}{CHO})_p-\text{(P)}$$

$$n = 0\text{-}3,\ m = 0\text{-}3,\ p = 0\text{-}3;\ x = 0\text{-}1;\ y = 1\text{-}6;\ \text{(P)} = \underset{O}{\overset{\|}{P}}(OH)_2,\ \underset{O}{\overset{\|}{P}}OH$$

Nevertheless some phosphorylation products described in the patent [645] exhibited a very high activity against scaling on metal surfaces by calcium sulphate, calcium carbonate and barium sulphate which made them particularly valuable in the oil production industry to protect the surfaces of pumps, tubes, pipes, valves, tanks and other metal parts. The effective concentrations of phosphorylated alcohols varied across the range from 0.5 up to 250 parts per million at pH 7.8–8.3. In cooling systems the required concentrations of alkyl phosphoric acids was found to be lower than in boilers.

Asphalts treated with phosphoric anhydride, which induced cross-linking, (Section 6.3) were found to be very good materials for controlling seepage in irrigation canals and water supply systems [646, 647]. Such asphalts were also very efficient in preventing pollution of ground water, control of industrial waste, the coating of settlement and filtration reservoirs of water supplies and suppressing the natural erosion of farm lands and beaches [646, 647].

2 OXIDATION

Phosphoric anhydride is used as a component of oxidizing mixtures in organic synthesis, the typical examples can be found in the publications [648–651] and in Chapter 3 of this book. A general method for the oxidation of alcohol to aldehydes and ketones was proposed by Taber and co-workers [649].

Unsaturated β-hydroxyesters were treated by phosphoric anhydride in DMSO at room temperature and the following reaction with triethylamine in dichloromethane gave the corresponding keto esters (Scheme 284).

$$CH_3CH = CH(CH_2)_2\underset{\underset{OH}{|}}{CH}CH_2CO_2Me \quad \xrightarrow{\begin{array}{c}1.\ P_4O_{10}/DMSO\\2.\ Et_3N(CH_2Cl_2)\end{array}}$$

$$\longrightarrow CH_3CH = CH(CH_2)_2\underset{\underset{O}{\|}}{C}\,CH_2CO_2Me$$

(284)

The efficient selective oxidation of sugars into the corresponding keto-forms by phosphoric anhydride–dimethysulphoxide or phosphoric anhydride–dimethylformamide mixtures is presented in Chapter 5.

A mixture of phosphoric anhydride and chlorine oxidized dichloroaryl phosphines into aryl phosphonic dichlorides in a highly exothermic process (exothermic effect up to 150°C) [648] (Scheme 285). The following esterification of the latter in the presence of a base resulted in the formation of the corresponding phosphonic acids (Scheme 151a) that are said to be of higher thermostability than phosphoric acids derivatives [648].

$$6\,ArPCl_2 + P_4O_{10} + 6\,Cl_2 \xrightarrow{\text{No solvent}} 6\,ArP(O)Cl_2 + 4\,POCl_3$$

$$ArP(O)Cl_2 + 2\,ROH + C_5H_5N \longrightarrow ArPO(OR)_2 + 2\,C_5H_5N \cdot HCl$$

$$Ar = Ph,\ C_6H_4Me - p,\ C_6H_4 - Cl - p;\ R = CH_2\underset{\underset{Me}{|}}{C} = CH_2$$

(285)

The products of this reaction were used in polymerization and copolymerization processes [648].

Peroxyphosphoric acid, $(HO)_2P(O)OOH$, which may be prepared from phosphoric anhydride and hydrogen peroxide in acetonitrile with cooling (Section 6.3) is a sufficiently powerful oxidizing agent to convert Mn^{2+} to permanganate.

Oxidation of methacrylaldehyde to methacrylic acid is catalyzed by a sponge of vanadium oxide–molybdenum oxide–phosphoric anhydride mixture [652] and the dioxygen oxidation of n-butane to maleic anhydride is promoted by titania-vanadium oxide–phosphoric anhydride catalyst [653].

3 PREPARATION AND APPLICATION OF INORGANIC COMPOUNDS DERIVED FROM PHOSPHORIC ANHYDRIDE

3.1 ORTHO PHOSPHORIC ACID AND INORGANIC PHOSPHATES

Phosphoric anhydride has many important uses and applications, not just in organic chemistry but also in inorganic chemistry. The major reactions of phosphoric anhydride, P_4O_{10}, with water, halogen compounds, nitrogen compounds, oxyacids, and with metal oxides, hydroxides, carbonates and sulphates yield many valuable commercial products.

Phosphoric anhydride is an important intermediate in the industrial production of phosphoric acid $P_4 \rightarrow P_4O_{10} \rightarrow H_3PO_4$. Nearly 80% of the phosphorus produced is utilized by the orthophosphoric acid industry [654]. In the thermal manufacturing method of phosphoric acid, phosphorus is oxidized in the stream of oxygen which is a highly exothermic process (ΔH 3 MJ/mol of P_4O_{10}). The following hydration of phosphoric anhydride is significantly less exothermic and it is usually carried out in an excess of recycled phosphoric acid [654]; modern technological processes are designed to include heat recovery. The thermal phosphoric acid produced is usually marketed in several concentrations — 75% H_3PO_4 (54% P_2O_5), 86% H_3PO_4 (62% P_2O_5), H_3PO_4 of high purity and polyphosphoric acid with a range of concentrations from 76% up to 84% (the so-called '116% H_3PO_4') [654].

The traditional methods of utilizing phosphorus for the production of phosphates and fertilizers are multistep processes (Scheme 286) that include phosphorus oxidation into phosphoric anhydride, its hydration into phosphoric acid, phosphoric acid neutralization followed by drying the corresponding salts (Scheme 286). From an energy-saving point of view the more efficient method avoids the production of phosphoric acid by the direct interaction between the various compounds and the P_4O_{10} fumes formed in phosphorus burning [655] (Scheme 287).

$$P_4 \xrightarrow{\ O_2\ } P_4O_{10} \xrightarrow{\ H_2O\ } H_3PO_4 \xrightarrow{\quad NH_3\quad} (NH_4)_3PO_4 \tag{286}$$

$$P_4 \xrightarrow{\ O_2\ } \underset{\text{fumes}}{P_4O_{10}} \xrightarrow{\ NH_3 + H_2O\ } (NH_4)_3PO_4 \tag{287}$$

The latter process utilizes the phosphorus burning energy evolved without any extra energy demand. This method has been used also for the production of potassium phosphate [655, 656].

Orthophosphoric acid which is produced by hydration of phosphoric anhydride (Chapter 3, Section 1.1) is one of the most widely used inorganic compounds [657, 658].

Following hydrolysis of purified phosphoric anhydride to metaphosphates and polyphosphoric acids, further addition of water gives pyrophosphates and eventually 'Food Grade' orthophosphoric acid. The latter, as well as being a key constituent of cola drinks, is converted to the weakly acidic sodium dihydrogen phosphate for use with the even more weakly acidic sodium hydrogen carbonate as a baking powder. Alkali phosphates are also used for the production of many buffer reagents [660] and for sequestering agents [661]. The sequestering power of tripolyphosphate towards calcium ions is better than most other polyanions and approaches that of EDTA. However, the largest use of phosphoric acid is as a key ingredient of fertilizers. This grade acid is manufactured by the 'wet' process from the reaction of naturally occurring apatites (in sedimentary rocks) with sulphuric acid (Scheme 288).

$$Ca_{10}(PO_4)_6F_2 + 10\,H_2SO_4 + 20\,H_2O \longrightarrow 6\,H_3PO_4 + 2\,HF + 10\,CaSO_4 \cdot 2\,H_2O$$

$$(288)$$

According to [662] about 80% of the phosphoric acid produced in the USA each year is used in the manufacture of fertilizers with the remainder being used for detergent additives, cleaners, insecticide production, and cattle feed additives.

Metal derivatives can often be prepared by disproportionation. The salts can also be prepared by heating the monobasic orthophosphates with a small excess of phosphoric acid above 300°C. Examples of metaphosphates include those of aluminium, barium [658], cobalt(II), iron(II), magnesium, manganese(II), nickel(II) cadmium, copper(II), and zinc [659]. Mercury(II), calcium, strontium, barium, and lead(II) give less well-defined products.

3.2 REACTIONS WITH HYDROGEN PEROXIDE AND OXYGEN: PEROXYPHOSPHORIC ACIDS

Permonophosphoric acid was first described in 1910 as arising from a very violent reaction between ortho phosphoric acid and hydrogen peroxide [663]. It is a very powerful oxidizing agent and in fact it is sufficiently powerful to convert Mn^{2+} to permanganate in the cold (Scheme 289a).

$$Mn^{2+} + 4\,(HO)\,P\,(O)\,OOH \longrightarrow Mn\,O_4^- + 4\,H_3\,P\,O_4 \qquad (289a)$$

Later studies showed that acetonitrile can be used as a diluent for its preparation [664]. Thus in this solvent with salt–ice cooling, phosphoric anhydride reacted over 4 h with 30% hydrogen peroxide to produce permonophosphoric acid, $(HO)_2P(O)OOH$ (Scheme 289b).

$$30\%\ H_2O_2 + P_4O_{10} \xrightarrow[\text{salt/ice cooling}]{CH_3CN} (HO)_2\,P(O)OOH \qquad (289b)$$

This reagent has been found to be useful for the Baeyer–Villiger oxidation of aldehydes avoiding any oxidation to the carboxylic acid. A patent [665] has been taken out for its use in the conversion of 3-benzocyclobutene aldehyde to 3-hydroxybenzocyclobutene which was achieved with 70% yield. Passing phosphoric anhydride vapour through a discharge tube produced elemental phosphorus and a violet coloured compound [666], which was shown to have oxidizing properties and it was suggested that an oxygen atom was inserted into a P–O–P bond of P_4O_{10} to give a P-O-O-P group. Treatment of the violet compound with water (4 equiv. H_2O) produced pyrophosphoric acids $H_4P_2O_8$ and $H_4P_2O_7$ and it was proposed that the compound has the structure shown in [667] (Scheme 290).

$$\xrightarrow{\text{H}_2\text{O}} \quad H_4P_2O_8 \quad + \quad H_4P_2O_7 \tag{290}$$

The overall process of interaction between phosphoric anhydride and the aqueous solution of hydrogen peroxide is shown in the following Scheme 290a.

3.3 REACTIONS WITH HALOGEN COMPOUNDS, NITROGEN OXIDES AND OXYACIDS

Halogen compounds

Phosphoric anhydride reacts with a variety of metal halides and hydrogen halides to yield mainly phosphoryl halides [668] (Scheme 291). The reaction with metal halides usually requires heating to 350°C or higher.

Phosphoric anhydride also reacts with dry hydrogen fluoride in a similar way to give phosphoryl fluoride and, for this reason, phosphoric anhydride cannot be used to dry hydrogen fluoride (Scheme 292).

Phosphoryl fluoride can also be made by the slow addition of fluorosulphonic acid to excess phosphoric anhydride. In the presence of excess anhydride 80% phosphoryltrifluoride can be isolated [669] (Scheme 293).

Pyrosulphuryl fluoride can be prepared by the use of excess fluorosulphonic acid above 55°C [670]. Phosphoric anhydride if added to 40% hydrofluoric acid, or heated to 135°C with ammonium fluoride, forms a series of fluoro-phosphoric acids [671] (Scheme 294).

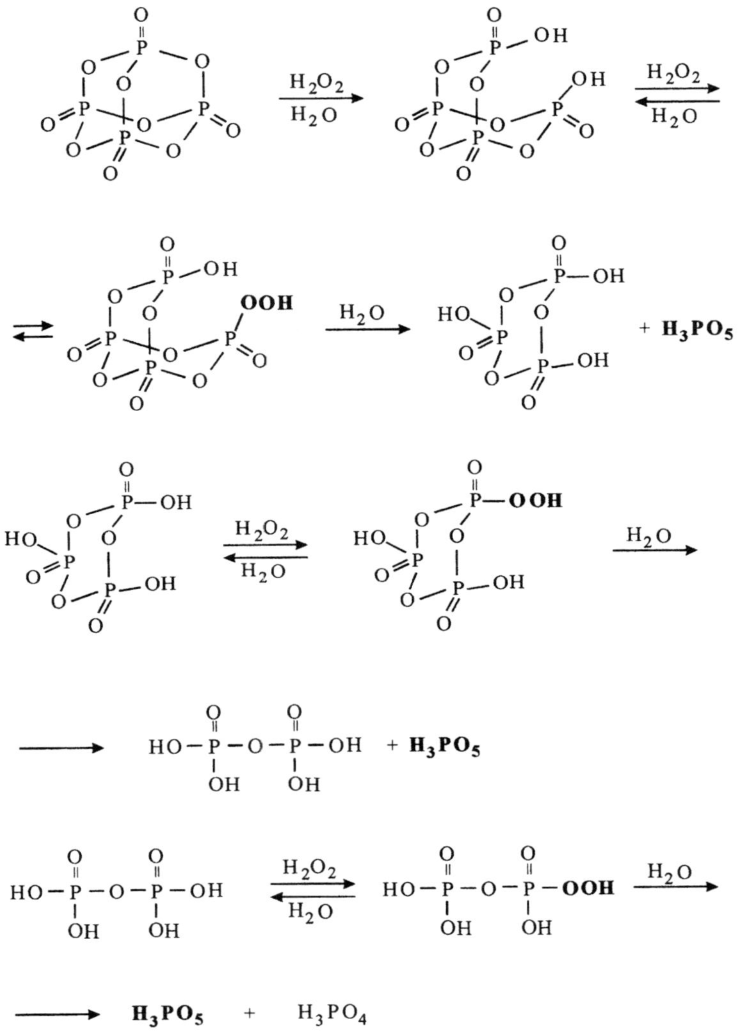

(290a)

$$3\,MX + P_4O_{10} \xrightarrow{\text{heat}} 3\,MPO_3 + POX_3 \quad (291)$$

$$3\,HF + P_4O_{10} \longrightarrow 3\,HPO_3 + POF_3 \quad (292)$$

$$HSO_3F \xrightarrow{P_4O_{10}} F_3PO + S_2O_3F \quad (293)$$

$$40\%\,HF \xrightarrow{\;P_4O_{10}\;} HPF_6 + HPO_2F_2 + H_2PO_3F \qquad (294)$$

Phosphoric anhydride also reacts with ammonium fluoride in ethanol solution to give 70% yield of ammonium difluorophosphate from which the potassium and cesium salts can also be prepared (Scheme 295).

$$NH_4F + P_4O_{10} \xrightarrow{\;\text{ethanol}\;} F_2P(O)ONH_4 \qquad (295)$$

Phosphoric anhydride also readily absorbs moist hydrogen chloride gas to give phosphorus oxychloride and metaphosphoric acid. The reaction is very slow if all traces of water are omitted [672] (Scheme 296).

$$P_4O_{10} + 3\,HCl \longrightarrow P(O)Cl_3 + 3\,HPO_3 \qquad (296)$$

Heating phosphoric anhydride with fluorite (CaF_2) at 500°C produces phosphoryltrifluoride as the major volatile product. In the presence of iron, some phosphorus trifluoride is also formed. Mixed phosphorus halides are produced when calcium fluoride and sodium chloride are heated together at 350°C [673].

Nitrogen oxides

The reaction of phosphoric anhydride with dinitrogen trioxide occurs slowly at ambient temperatures to give an adduct $P_4O_{10}\cdot4NO$, the same adduct being obtained using N_2O_4 at 250°C [674, 675]. It is also possible to obtain $P_8O_{21}\cdot2NO$ [676].

Oxyacids

Phosphoric anhydride dehydrates oxyacids such as sulphuric, nitric and perchloric acids forming the corresponding acid anhydrides [677] (Scheme 297).

$$
\begin{aligned}
P_4O_{10} + H_2SO_4 &\longrightarrow SO_3 + \text{polyphosphoric acid} \\
P_4O_{10} + HNO_3 &\longrightarrow N_2O_5 + \text{polyphosphoric acids} \qquad (297)\\
P_4O_{10} + HClO_4 &\longrightarrow Cl_2O_7 + \text{polyphosphoric acids}
\end{aligned}
$$

A patent for the latter process for the manufacture of chlorine VII oxide recommends carrying out the reaction between 5 and -5°C in the presence of some chloroform or dichloromethane to decrease the danger of explosion [678].

The reactions are undoubtably driven by the formation of stable polyphosphates with the generation of very good leaving groups in the presence of powerful Brönsted acids and with the generation of more volatile products, which can be removed from the less volatile reactants.

Phosphoric anhydride forms a series of solid solutions with arsenic anhydride from which oxides of the type $AsPO_5$ are formed [679].

3.4 REACTION WITH METAL OXIDES, HYDROXIDES, CARBONATES, SULPHATES AND NITRIDES

Both phosphoric anhydride and orthophosphoric acid are used widely for the preparation of phosphates and pyrophosphates (also called diphosphates and dipolyphosphates) from oxides and other salts. Orthophosphoric acid is more commonly used in hydrothermal reactions where solutions of starting materials are heated in autoclaves at high temperatures. The reactions of the anhydride are often carried out in the solid phase and occasionally in organic solvents. They tend to be more complicated in that more phosphate intermediates are involved than in the preparation of phosphates from phosphoric acid.

In the presence of water phosphoric anhydride reacts with all basic oxides and, oxides such as Al_2O_3 [680], SiO_2 [681], and boric oxide [682]. Ferric oxide, Fe_2O_3, produces a series of metaphosphates, phosphates and pyrophosphates; their interconversions occurring at different temperatures [683, 684] (Scheme 298).

$$P_4O_{10} + Fe_2O_3 \xrightarrow{325^\circ} Fe(PO_3)_3 \xrightarrow[P_4O_{10}]{650^\circ} FePO_4 \xrightarrow[Fe(PO_3)_3]{800^\circ} Fe_4(P_2O_7)_3 \quad (298)$$

In the reaction of lime with phosphoric anhydride one drop of water into the solid mixture induces such a violent reaction that the mixture bursts into flames. In the case of phosphoric anhydride and sodium carbonate local heating will also initiate a very vigorous process, the reaction mixture becoming so hot as to vapourize the phosphoric anhydride and produce phosphate glasses. The reaction of phosphoric anhydride with calcium carbonate has been studied over the temperature range 25–1000°C by means of thermal analysis [685]. A similar study has been carried out using calcium sulphate [686]. Under anhydrous conditions heat is required, fusion with basic oxides giving solid phosphates the structures of which depend on the reaction conditions and the ratio of the reactants. For sodium oxide a continuous series of vitreous sodium phosphates from P_4O_{10} through to Na_2O has been studied. Reaction of phosphoric anhydride vapour with silica also produces a sequence of silicon phosphates [687]. Reaction with two equivalents of sodium carbonate produces sodium phosphate with the liberation of carbon dioxide [688], whereas, with the basic carbonate in the presence of water, polyphosphate, $Na_4H_2P_4O_{13}$, is produced. It reacts with basic oxides such as zinc oxide in water to give a hydrated zinc hydrogen phosphate, $Zn_3(PO_3(OH))_3 \cdot 3H_2O$ [689]. Bismuth oxide and P_4O_{10} heated together in air gives a fluorescent phosphate $Bi_{18.7}P_{13}O_{31.3}$ [690] (Scheme 299).

$$Bi_2O_3 + P_4O_{10} \xrightarrow[\text{air}]{650°} \underset{\text{fluorescent compound}}{Bi_{18.7}O_3P_{13}O_{31}.} \qquad (299)$$

Neodinium metaphosphate $Nd(PO_3)_3$ has been made by the reaction of the oxide Nd_2O_3 with P_4O_{10} in the presence of alkali metal carbonate followed by heating the resultant ultraphosphate in vacuo [691] (Scheme 300).

$$Nd_2O_3 + P_4O_{10} \xrightarrow{Li_2/Na_2/K_2CO_3} MNd(PO_3)_4 + Nd\ ultraphosphate$$

$$\downarrow \text{heat in vacuo}$$

$$Nd(PO_3)_3$$

$$(300)$$

In the case of a similar reaction using titanium oxide, $K_3Ti_3P_5O_{20}$ crystallized spontaneously from the melt [692]. Related alkaline earth–transion metal pyrophosphates have been prepared incorporating iron, cobalt and nickel oxides [693]. Vanadium oxide has been converted from the appropriate oxides to $KV_3P_4O_{16}$ [694] and to $SrV_2(P_2O_7)_2$ [695] and also to $Zn_2VO(PO_4)$ [696]. Titania (TiO_2) was used to make $BaTi_2(P_2O_7)_2$ [697] and molybdenum oxide to make $Cs_4Mo_6P_{10}O_{38}$ [698] (Scheme 301).

$$MoO_3 + P_4O_{10} \xrightarrow[950°C]{Cs_2CO_3} Cs_4Mo_6P_{10}O_{38} \qquad (301)$$

Palladium oxide undergoes an interesting reaction with P_4O_{10} at 700°C in air giving rhombic crystals of PdP_2O_6 [699, 700] (Scheme 302).

$$PdO + P_4O_{10} \xrightarrow{700°} PdP_2O_6 \qquad (302)$$

There are examples of compounds with mixed metal cations which incorporate neodinium and also vanadium [701] (Schemes 303, 304).

$$Nd_2O_3 + P_4O_{10} + ZnO + Na_2O \longrightarrow Na_{1+x}Zn_{2-x}Nd_x(PO_4)_3 \qquad (303)$$

$$Cd_2P_2O_7 + 2Co_2P_2O_7 \xrightarrow{P_4O_{10}} 2CdCuP_2O_7 \qquad (304)$$

A variety of mixed pyrophosphates centred around copper, cobalt and cadmium have been prepared using phosphoric anhydride and their catalytic and physical properties studied.

Cobalt pyrophosphates that are made from cobalt nitrate and ammonium hydrogen phosphate [703] have been converted to various mixed phosphates.

For example intimate mixtures of cadmium diphosphate and cobalt diphosphate were heated for various periods in stages at 873K, 1173K and then with phosphoric anhydride at 1373K before cooling [704]. A byproduct $Co_3CdP_4O_{12}$ was also formed. The mixed diphosphates were studied by neutron diffraction [705] and their catalytic properties assessed [706] (Scheme 305).

$$Cu_2P_2O_7 + 2Co_2P_2O_7 \xrightarrow{\ P_4O_{10}\ } 2CoCuP_2O_7 \tag{305}$$

The corresponding reaction between copper pyrophosphate and cobalt diphosphate was also studied [707] and two additional cobalt-rich products characterized.

Phosphoric anhydride reacts with lithium nitride upon heating to give Li_3PN_2O [702].

3.5 LIGAND MODIFICATION VIA DEHYDRATION

The hexagonal form of phosphoric anhydride is the most efficient dehydrating agent. It is the normal commercial product being produced when phosphorus is burned in oxygen and condensed on a cold surface. It is easy to use and can be applied to the chemical elimination of water from alcohols, amides as well as for the preparation of anhydrous salts.

Conversion of hydroxylic ligands to alkenes Applications involving the dehydration of hydroxylic compounds by P_4O_{10} may be extended to the preparation of inorganic complexes of unsaturated esters. Thus rhenium and manganese tricarbonyl complexes of cyclopentadienyl-3-hydroxybutyrate esters are dehydrated in benzene by phosphoric anhydride to form stable Re and Mn complexes of butenoates, which are sufficiently stable to be purified by chromatography on alumina [708] (Scheme 306).

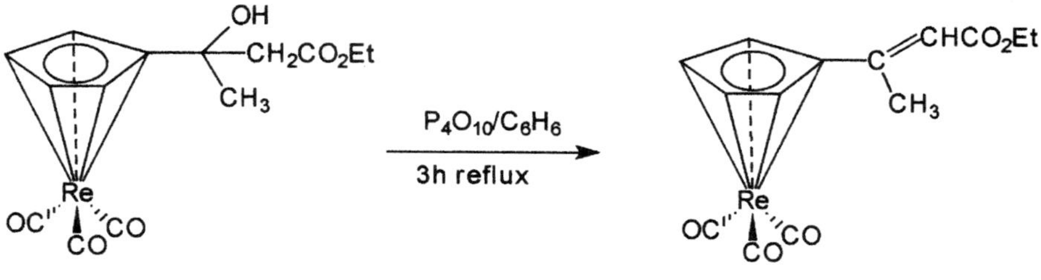

$$\tag{306}$$

In a similar manner chromium 3-phenyl-3-hydroxypropionate may be converted to the corresponding propenoate derivative (Scheme 307).

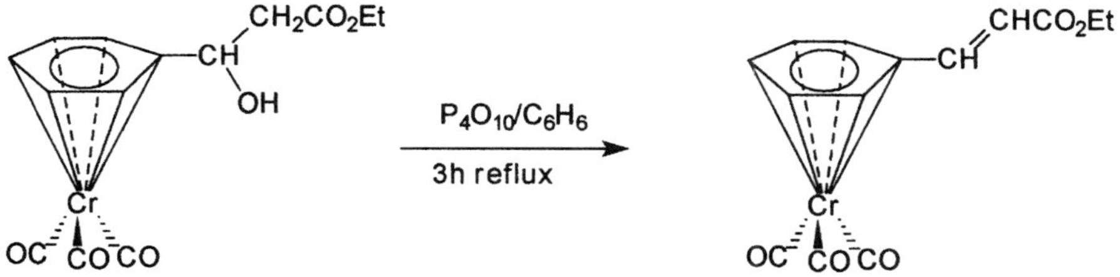

$$\text{(307)}$$

Conversion of amide to nitrile complexes   Phosphoric anhydride has also been successfully used for the synthesis of nitrile adducts and complexes. Examples include the dehydration of tris(trifluoromethyl) boroacetamide hydrate by P_4O_{10} in boiling acetonitrile (12 h) to give tris(trifluoromethyl) boroacetonitrile which was purified by vacuum sublimation [709]. Carboxylic acids amides and diamides [710, 711] are converted by P_4O_{10} at 170°C in liquid paraffin to the mono and dinitrile, respectively, each of which react with rheniecene hydride to produce stable complexes $Cp_2ReCHCH(CN)$ and $Cp_2ReC(CN)CH(CN)$, respectively (Scheme 308), that were purified by chromatography on basic alumina.

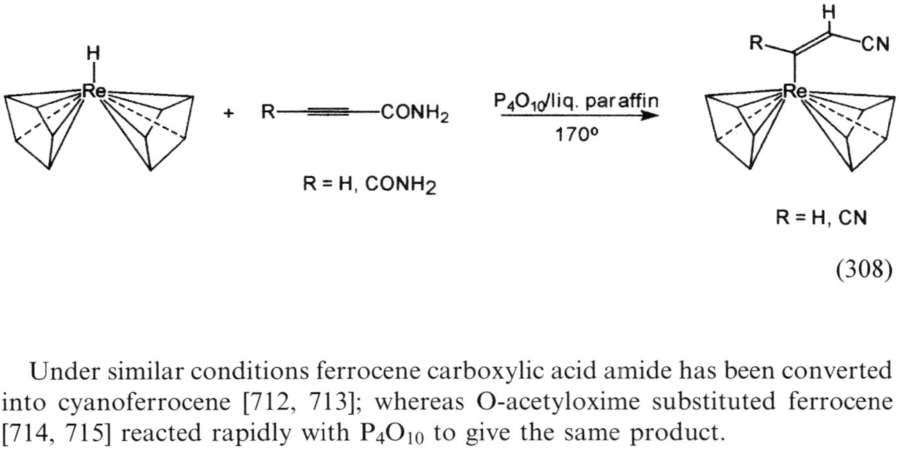

$$\text{(308)}$$

Under similar conditions ferrocene carboxylic acid amide has been converted into cyanoferrocene [712, 713]; whereas O-acetyloxime substituted ferrocene [714, 715] reacted rapidly with P_4O_{10} to give the same product.

Synthesis of other organometallic compounds via dehydration processes   2,4,6-Tris(ferrocenyl)boroxin was prepared by the action of phosphoric anhydride on the corresponding ferrocenyldihydroxyborane (Scheme 309).

$$P_4O_{10} + (C_5H_5)Fe[C_5H_4B(OH)_2] \longrightarrow [(C_5H_5)Fe(C_5H_4B(O)]_3 \qquad \text{(309)}$$

3.6 DEHYDRATION OF INORGANIC HYDRATES

Phosphoric anhydride is a widely used powerful, yet relatively safe, drying agent. Indeed it is one of the most efficient at temperatures below 100°C. Examples include its use neat for the preparation of anhydrous lithium formate from its monohydrate at room temperature [716], to prepare anhydrous lithium permanganate from its trihydrate [717]; and for the preparation of the dihydrate of tellurium dioxide–pyridine adduct from its decahydrate [718].

3.7 INDUSTRIAL INORGANIC APPLICATIONS OF PHOSPHORIC ANHYDRIDE

Use as a catalyst The role of phosphoric anhydride in the production of catalysts is vast and the petroleum industry has made extensive use of its ability to improve selectivity and conversion efficiency. There are patents describing the use of phosphoric anhydride as a catalyst ingredient or as a promoter for the vapour-phase catalytic syntheses of maleic anhydride, acrolein, acrylic and methacrylic acids, aldehydes, nitriles, butadiene and others [719–722].

Most are vapour phase oxidations. Thus dioxygen oxidation of methacrylaldehyde to methacrylic acid is catalyzed by a sponge of vanadium oxide–molybdenum oxide–phosphoric anhydride mixture [723] and the dioxygen oxidation of n-butane to maleic anhydride is promoted by titania–vanadium oxide–phosphoric anhydride catalyst [724].

Phosphoric anhydride exhibited pronounced catalytic properties in the polymerization processes of olefins [843].

Phosphate glasses Phosphate glasses are formed in violent reactions between phosphoric anhydride, caustic soda and calcium oxide [725]. Glasses are also the products of the reaction of calcium phosphate with the anhydride at 1000°C [726]. Phosphate glass has been used to improve the transparency of china ware [727].

Silicate glasses Phosphoric anhydride has been widely used to aid the formation of alkali metal silicate glasses at 850°C e.g. the preparation of yttrium sodium silicate $Na_3YSi_6O_{15}$ [728]. These glasses have various industrial applications (see below).

- **Detergents and soaps** A product from ammonia and P_4O_{10} (Victamide) has been marketed as a heavy duty liquid anionic detergent [729], whereas a fused glass of phosphoric anhydride with magnesium oxide and sodium oxide with or without potassium oxide has been used in specialist soaps [730]. Sanitizing properties are added to soap and tetrasodium pyrophosphate by the addition of phosphoric anhydride and silver oxide [731].

- **Metal cleaners** Phosphates and polyphosphates are used widely in metal cleaners. Fused vitreous products for this application are based on formulae $Na_2O \cdot SiO_2 \cdot P_2O_5$ combined in a wide range of proportions [732, 733]. Boric acid when reacted with phosphoric anhydride in the presence of sodium phosphate gives a water soluble cleaning agent [734].
- **Surface properties and sealants** A lubricant and grease additive, which improves wear resistance, was prepared from a mixture of P_4O_{10}, P_4S_{10}, amine, sulphur and C_2–C_{32} alkene [735].

  Alumina rods have been sealed using a mixture of phosphoric anhydride and aqueous HF. The strength of the seal was found to improve as the temperature was raised from 800 to 1,220°C [736]. Ammonium poly-phosphate is added to molding polyolefin resins to impart bleeding resistance and flameproofing. The methods of production involve heating phosphoric anhydride with ammonium phosphate in the presence of ammonia gas [737, 738].

Asphalt Among modifiers and cross-linking agents of asphalt such as ferric chloride, red phosphorus, phosphorus sulphide, phosphorus sesquisulphide, phosphoric acid, phosphoric thioanhydride and phosphoric anhydride, the last one possesses a number of advantages that have made it the most suitable [739, 740]. Phosphoric anhydride is added for catalytic air blowing in quantities from 0.1 up to 3.0% depending upon the asphalt characteristics desired. The authors [739, 740] did not discuss the actual mechanism of phosphoric anhydride insertion in asphalt, nor the structure of the products formed but phosphoric acid could not be removed from the resulting asphalt by water extraction which is not the case with ferric chloride. The section of the article [739] devoted to the properties of modified asphalt was meaningfully headed: 'Phosphorus Pentoxides Asphalts Have Better Temperature Susceptibility and Increased Weathering Characteristics'. The most valuable property of asphalt modified by phosphoric anhydride is its high stability on heating which results in better resistance to flow in summer, more pronounced adhesiveness as well as low brittleness at low winter temperatures due to their higher viscosity and temperature stability. Phosphoric anhydride asphalts have a prolonged life when used as a basis for tank paints. Under conditions of heavy traffic such asphalts showed much better resistance than ordinary asphalts or asphalts incorporating rubber [739].

4 APPLICATION OF ORGANIC PHOSPHATES DERIVED FROM PHOSPHORIC ANHYDRIDE

There are numerous examples of efficient phosphorylation of alcohols by neat phosphoric anhydride or by mixtures P_4O_{10}/H_2O, P_4O_{10}/H_3PO_4, $P_4O_{10}/H_2O/$

H_3PO_4, P_4O_{10}/polyphosphoric acid, polyphosphoric acid and others that have been presented in patents, scientific publications, industrial bulletins and brochures as well as textbooks. It is worth mentioning that in a few references, for example [741] one can find application limitations for monoalkyl (aryl) and dialkyl (aryl) phosphoric acids due to their corrosive effect.

There are a few contradictory papers devoted to the applications of phosphates synthesized with phosphoric anhydride. For example some of the arguments against phosphates are their low water solubility, poor resistance to hard water, and so on [742] but the conclusion in the same publication stated that phosphates are *'the unique class of anionic surfactants'* [742]. From this critical point of view the authors gave a systematic analysis of anionic surfactants. It was pointed out that the most valuable properties originate from the presence of the dihydroxyphosphoryl group. Among the properties that make alkyl, aryl and arylalkyl phosphates and their mixtures widely useful are heat, light and mechanical stability, film clarity and buffer action [742].

The phosphoric acids alkyl (aryl) esters and their mixtures have been produced industrially for many decades from phosphoric anhydride and the corresponding alcohols or trialkyl (aryl) phosphates have been used on a broad scale since the early 1930s [107]. Later studies have been focused predominantly on improving the earlier methods [743, 744 and references therein]. The esters can vary from methyl esters to long-chain esters, their thermal stability being considered to be the only limitation for their application.

4.1 SURFACTANTS, DETERGENTS AND RELATED APPLICATIONS

An in-depth study of the synthesis and surface activity of aminophosphates was presented fairly recently by Tsubone and co-workers [127] (Chapter 3, Section 1.2) and other researchers [745]. These products are of particular interest for industrial application as polyfunctional surfactants that possess two hydrophobic centres (the amino and dihydroxyphosphoryl groups) separated by a relatively short chained bridge [-N-CH$_2$-CH$_2$-O-P(O)(OH)$_2$]. Three compounds were investigated N-dodecyl-N-methylaminoethyl phosphate, N-tetradecyl-N-methylaminoethyl phosphate and N-hexadecyl-N-methylaminoethyl phosphate. On the basis of surface tension–concentration plots and the presence of the distinctive break points in the range of concentrations 10^{-4}–5×10^{-3} (moles/l), it was concluded that they form micelles in aqueous solutions. The break points for phosphates derived from higher molecular weight alcohols were at the lowest concentrations. The logarithm plot of cmc against the number of carbon atoms in the hydrophobic chain had an inverse relationship. Surface tension (σ) was determined for $C_{12}H_{25}(CH_3)NCH_2CH_2OP(O)(OH)_2$ and found to be 39 dyne/cm^2; this characteristic value did not vary much for the other two phosphates [745].

The occupation area at the surface was found to be the highest for $C_{16}H_{33}N(CH_3)$ $CH_2CH_2OP(O)(OH)_2$ which agrees well with the theory of surface active compounds. The foaming activity of the above-mentioned products [745] decreased across the hydrophobic substituents $C_{12}H_{25} > C_{14}H_{29} > C_{16}H_{33}$ in this order. This was explained by the excellent solubility of the former compound in water. Its foaming activity was not influenced by acidity in the pH range 4.0 to 8.0. In contrast the other two compounds exhibited maximum activity only in a very narrow range of pH. The effect [745] was attributed to the zwitterionic character of N-dodecyl-N-methylaminoethyl phosphoric acid in the media with pH $c.$ 5.0–9.0.

The textile industry is one of the largest consumers of organophosphates produced from phosphoric anhydride [746–749]. *The Handbook of Industrial Surfactants* [750] contains a list of dozens of producers and many dozens of entries devoted to acidic organic phosphates which are assigned specific properties and produced industrially from phosphoric anhydride, polyphosphoric acid or their derivatives. Among the producers is Burlington Chemical Co (USA) which manufactures surfactants (BurcoFacTM) that are complex phosphate esters — mixtures of mono- and di-alkyl phosphoric acids. The textile industry takes advantage of the high surface activity, fire-retarding ability, alkali and hard water stability, lubricating and other related properties of organic phosphates [102, 751–766].

A variety of alcohol and alcohol ethoxylate hydrophobes are phosphorylated in Burlington Chemical Co utilizing phosphoric anhydride or polyphosphoric acid to provide surfactants for textiles, i.e. hard surface cleaners, metal working fluids and lubricants that have many of the excellent detergent properties of non-ionic surfactants with the electrolyte stability of anionics.

The earliest important commercial applications of organic phosphates derived from phosphoric anhydride and natural vegetable oils or animal fats in textile, leather and pharmaceutical industries were emphasized by Bertsch [767]. The mixed esters of phosphoric acid and ricinoleic or oleic acid were produced in the reactions of phosphoric anhydride with the corresponding acid in the presence of either acetic anhydride or acetyl chloride; other conditions were unspecified. After washing out phosphoric acid the products were neutralized by sodium carbonate. The resulting sodium salts were readily soluble in water and showed to be excellent washing agents. Unlike the products from the same organic acids and sulphuric acid, the phosphorylated compounds possessed a number of advantages, including higher purity and pale colour [767, 767a]. The last point is an important consideration in the application of some valuable but deeply coloured products.

Individual mono(polyfluoroalkyl) phosphoric acids and bis(polyfluoroalkyl) phosphoric acids or their mixtures produced in the reaction of phosphoric anhydride with polyfluoroalcohols (Chapter 3, Section 1.2) were determined to be efficient surface active agents and dispersing agents in the processes of tetrafluoroethylene and chlorotrifluoroethylene polymerization in water media. It is pointed out that polyfluoroalkylphosphates did not cause termination of chains in the step chains propagation as is specific for conventional dispersing agents.

Detergents Organic phosphates have been found to be among the most efficient detergents. Particularly active compounds are the products of condensation of lauric acid and polyethylene oxide or polypropylene oxide phosphorylated by phosphoric anhydride [124, 766, 768]. Di(2-ethylhexyl) phosphate, the synthesis of which is described in Chapter 3, Section 1.2, is widely used in industry as a surface active compound, either in its pure form or in the presence of mineral surfactants. Some salts of organic phosphates are used as tablet detergents with good disintegration properties [767a, 767b].

Efficient *fabric softeners* are produced from long-chain alkanols and phosphoric anhydride in the form of monoalkyl and dialkyl phosphoric acid mixtures [770, 771].

Phosphates containing cyclic substituents such as cyclooxyneopentyl phosphate and acyl oxycyclobutane phosphate exhibited very high stabilities under hydrolytic conditions and at high temperature, retaining lubricating activity under extreme pressure and surface activity [772]. The phosphates were produced by the reaction of phosphoric anhydride with an alcohol or glycol ester at an elevated temperature in ethylene chloride. The dominating ingredients of the composition were the corresponding monoalkyl and dialkyl phosphates that could be used without separation either in the form of free acids or as sodium, potassium, calcium or ammonium salts. In addition to the above-mentioned applications various products can be used as powerful *wetting and ore flotation agents.*

In the manufacture of fats and oils, acidic phosphates and particularly mixtures of mono- and disubstituted phosphoric acids, a conventional acid-base titration is used for the determination of the so-called 'acid value' [773]. It is expressed in milligrams of 1 N solution of KOH, which is required to neutralize the fatty or other acid contained in a gram of the sample. Acid value or acid number presents indirect information about product quality but does not specifically identify the product. Acid value is widely used in the quality control of surfactants including those possessing the hydroxyphosphoryl group.

It is possible to enhance the properties of surfactants by the use of formulations which include organic phosphates possessing anti-static and anti-corrosion properties as manufactured by the Burlington Chemical Co.

For textile finishing baths foaming is an undesirable process which can be successfully suppressed by water-insoluble esters of phosphoric acid [776]. For this purpose the preferable length of the alkyl chain was established to be C_{16}–C_{20} [777, 778]. In some cases this property of alkyl phosphates (stearylalkoxy phosphates) was aided by the presence of another component (a mineral oil) [778].

The mixture of products formed as shown in Scheme 309a were studied for *wetting activity* and *alkali tolerance* [122]. It was found that the alkali stability of alkyl phosphates became more pronounced with an increasing number of OCH_2CH_2 blocks in the substituents. Among alcohols with the same number of ethylene oxide fragments but different alkyl substituents, there was a higher alkali tolerance for compounds with a lower molecular weight. The wetting time increased in proportion to the chain length in the hydrophobic fragments. The presence of two or three ethylene oxide fragments in a molecule resulted in the fastest wetting time.

Eiseman and Schenk [779] proposed initiating the reactions of phosphoric anhydride with primary aliphatic alcohols (C_4–C_8) and alkoxyalkanols by hypophosphorus acid, phosphorus acid or their salts. The same alcohols can also be phosphorylated by polyphosphoric acid (115%, containing 1 mole of P_4O_{10}) [779] with the formation of a mixture of monoalkyl and dialkyl phosphoric acids that were claimed to be pronouncedly *alkali soluble* and alkali stable products.

Mansfield and Hill [780] patented a series of highly alkali soluble and alkali stable compounds produced in the reaction of polyphosphoric acid (82–85% content of P_4O_{10}) and non-ionic surface active compounds synthesized from primary amines and ethylene oxide (Scheme 309a).

$$R-NH_2 + xCH_2-CH_2 \longrightarrow R-N\underset{(CH_2CH_2O)_{x-y}H}{\overset{(CH_2CH_2O)_yH}{<}} \xrightarrow{P_4O_{10}\,/\,PPA}$$

$$(309a)$$

$$\longrightarrow R-N\underset{(CH_2CH_2O)_{x-y}\,\overset{O}{\underset{\|}{P}}(OH)_2}{\overset{(CH_2CH_2O)_y\,\overset{O}{\overset{\|}{P}}(OH)_2}{<}}$$

$$R = C_nH_{2n+1}\;;\quad n = 8\text{-}14\;;\quad x = 5\text{-}50\;;\quad y = 5\text{-}50\;;\quad 5 \leqslant x+y \leqslant 50$$

The preferred molar ratio of diol to polyphosphoric acid was between 1.0:1.2–4.0. The actual amount of polyphosphoric acid within this ratio was dependent upon the number of ethylene oxide fragments incorporated in the particular diol intermediate [780]. The authors [780] emphasized that in many cases the molar ratio of alcohol per mole of polyphosphoric acid had to be not less than 1:2, otherwise the process of phosphorylation could remain incomplete which resulted in a product with a lower alkali solubility and poorer alkali stability. The optimum ratio of the reagents for a reaction lasting 2–6 h was established to be 2.0 : 2.6. The best procedure consisted of the gradual adding of polyphosphoric acid to the alcohol at an elevated temperature and heating the reaction mixture at 80–110°C until the process completed, this was indicated by a stable level of cloud point determined in dilute sodium hydroxide solution. Reversing the order of charging the reaction flask unfavourably darkened the product. To avoid colour developing, small quantities of hydrogen peroxide were mixed with the alcohol prior to starting the reaction. Alternative decolourizing agents need not be oxidising agents but could be a reducing agent such as hydrazine hydrate.

Application of phosphoric acid as a phosphorylating agent of the alcohols instead of polyphosphoric acid under similar conditions gave a very low yield of the desired products.

In comparison with polyphosphoric acid, halogen-containing phosphorylating agents possess numerous drawbacks including the necessity to use glass-lined equipment because of the liberation of hydrogen chloride in large quantities [780].

The phosphorylated alcohols synthesized in the patent [780] exhibited a higher alkali solubility and alkali stability than the majority of well known nonphosphorylated nonionic surfactants.

The influence of acidity and the nature of the base upon the *foaming activity* of monododecyl phosphoric acid produced in the reaction of n-dodecanol with phosphoric anhydride under specific conditions have been studied [125]. It was established that the most efficient foaming can be achieved with the monoalkyl phosphate, with purity of 80% or higher. The sodium salt of this compound gave a larger volume of foam than the corresponding triethylamine salt. The optimum pH for foaming was about 7.0 and the monobasic salt of monododecyl phosphoric acid was found to be more active than its dibasic salt.

Dialkyl phosphoric ester synthesized from 2-hydroxyethylene methacrylate and phosphoric anhydride has been claimed to be an active agent for *reinforcing fibres* and *synthetic resins* [781].

Polyalkoxyalkyl phosphoric acids produced according to Scheme 30 (Chapter 3, Section 1.2) exhibit the distinctive features required of foaming and emulsifying agents, detergents and all-purpose cleaners [761, 762, 782–785]. For example Nunn [782] synthesized a surfactant from 1 mole of phosphoric anhydride and 4–9 moles of an alcohol consisting of at least one ethylene oxide fragment and an alkyl chain of 6–150 carbon atoms including such groups as phenols, fatty acids, fatty amines, sulphonamides, mercaptans and some others. An excess of alcohol was not required in the reaction, which was carried out at 110°C or below. The best phosphorylation conditions involved using inert solvents such as benzene, xylene, ether, pentane or hydrocarbons in various proportions. Contrary to the author's claim [782] ether media is not inert when brought into contact with phosphoric anhydride (Chapter 3, Section 1.3). According to the Ross–Miles foam test the most active products with the best foam stability were derived from phosphoric anhydride and dodecylphenol+18 ethyleneoxide fragments, tridecanol+10 ethyleneoxide fragments and nonylphenol+10 ethyleneoxide fragments. Several alcohols of the series possessed no surface activity.

The products of condensation of alcohols (C_{10}–C_{13}) with 10 or 18 moles of ethylene oxide phosphorylated by phosphoric anhydride have been patented [784, 786]. The most valuable feature of the products was sequestering water hardness by complex formation with magnesium and calcium cations.

For some particular applications an individual mono- or dialkyl phosphate is of higher activity than a mixture of the two. Moreover such mixtures often contain free orthophosphoric acid which is an undesirable factor.

Burlington Chemical Co strictly controls the ratio of monoesters to diesters of phosphoric acid to ensure consistency in wetting and emulsion properties. The products are manufactured as free acids, and neutral esters to produce potassium, sodium or ethanolamine salts depending upon the desired end use. The manufacturing process is based upon the reaction of phosphoric anhydride or polyphosphoric acid with β-alkoxy (C_9–C_{11}) ethanol, β-dodecyloxy ethanol, β-phenoxy ethanol, butyl cellosolve, 2-ethylhexanol and some other alcohols. BurcoFac^{TM} phosphate esters are among the most popular surfactants for textile open width fabric wet processing. They are also very useful in the dispersion of dyes and pigments in package-dyed yarn and beam type knit goods processing. BurcoFac^{TM} products find their way into scouring agents for bleaching and preparation, as wetting and dispersing agents in sulphur dyed cellulose, as dispersion agents for disperse dyeing polyester, and as the emulsification agent in latex polymer manufacturing, and dispersing agent in pigment manufacturing. Because the surfactants function well at most pH values, these products offer a versatile detergent/surfactant that can be used in all facets of textile wet processing and offering economies and formulation flexibility.

The *separation* of mono- and dialkyl phosphoric acids can be readily accomplished by partitioning the two phosphates between two immiscible liquids that are generally water and an organic solvent [787, 788]. Monoalkyl phosphoric acids are more soluble in water than dialkyl phosphates which have a high solubility in organic solvents such as diethyl ether, di-isopropyl ether and di-n-butyl ether. The most satisfactory separation is achieved when a monoalkyl phosphoric acid partition K is greater than 5 [788]. This method was efficiently applied to the separation of the corresponding mono- and di-butyl, amyl and octyl phosphates [788]. Another method of alkyl phosphoric acid separation is based on the different solubility of their calcium and barium salts [789]. Calcium and barium salts of monoalkyl phosphates are sparingly soluble in water, while the corresponding salts of dialkyl phosphoric acids have pronounced solubility in water.

Various *cosmetic products* are formulated with a high content of polyethoxyalkyl phosphates, alkyl phosphates and arylalkyl phosphates due to their low toxicity and good emulsifying activity in oily media [790–792]. Efficient detergents have been synthesized [125] by the reaction of phosphoric anhydride with alcohols (C_8–C_{12}) in the presence of water (Chapter 3, Section 1.2) which are inert towards the contact with human skin. There have been other similar synthetic approaches to the production of surfactants [793, 794].

Additives to liquid petroleum hydrocarbons, ranging from petrol through to heavy lubricating oils, were produced in the reaction of phosphoric anhydride with polyglycol ethers [797]. Formation of the corresponding monoalkyl and dialkyl phosphoric acids was complete within 1–5 h of heating the reaction mixture at 130–180°C. For better solubility in hydrocarbon liquids the addition of alkylamines of linear or iso-structure with the alkyl chain C_1–C_{24} has been recommended [797]. Some of the products were shown to be excellent rust inhibitors and suitable as fuel additives or lubricating oil additives for the protection of engines and also as agents for reducing the formation of sediment in fuel oils [797].

4.2 FIRE RETARDANTS

Organic phosphates have been widely studied and used as efficient flame retardants for many decades. Nevertheless in the past 20–30 years this class of flame retardant has gained a new impetus particularly with the invasion of synthetic polymeric materials in all aspects of technology and everyday life [636]. Organic phosphates are particularly effective flame retardants in textile and related industries dealing with cotton, wool, polyesters, polyurethanes, acrylic fibres and others due to inhibiting ignition and promoting char formation. One of the most attractive features of organophosphorus flame retardants (except direct activity) is their environmental friendliness in all respects: clean production, very low toxicity and minimal harmful volatile product formation in the burning process [636, 797a].

In this respect organic phosphates are particularly valuable flame retardants in isolated areas with a high risk of ignition such as airplanes and automobiles.

Surprisingly enough the phosphorylation of natural monomers by phosphoric anhydride has been investigated on a limited scale. Though, as can be seen in Chapter 4, phosphoric anhydride as well as polyphosphoric acids derived from it could have been the primary phosphorylating agents of organic compounds on the primitive Earth. A rare example of a study of the phosphorylation of a natural monomer by a mixture of phosphoric anhydride and phosphoric acid was updated recently [798]. Anhydrous D-glucose was phosphorylated by a mixture P_4O_{10}–H_3PO_4 in dioxane at 50°C. Probably the overall reaction rate was enhanced by the preliminary addition of phosphoric anhydride to D-glucose followed by phosphate condensation with ethylene oxide, which is known to be very reactive towards P_4O_{10} (Chapter 3, Section 1.4) and polyphosphoric acid. Though the phosphorylated glucose was not isolated [798] it was assumed to be the product of dehydration–phosphorylation, and built of five carbohydrate fragments with the dihydroxyphosphoryl group at a terminal structural block in an unspecified position.

Phosphorylated glucose in combination with another organophosphorus compound known as Pyrovatex CP, i.e. $HOCH_2NHC(O)CH_2CH_2P(O)(OMe)_2$, and a melamine derivative exhibited pronounced flame-retarding properties in cotton. Synergism of phosphorus–nitrogen in flame retardant was observed in this case as well as in many other studies.

The two-step process which included phosphorylation of hydroxyalkyl acrylate or methacrylate with phosphoric anhydride followed by blocking all three hydroxyl groups by ethylene oxide gave monomers for flame-retardant resin compositions [799]. The patent [799] states that any other ratio than 1.5 mole of alcohol per each phosphorus g-atom caused undesirable problems with the purification of the products. The optimum temperature of phosphorylation (30–40°C) was claimed to be one of the lowest reported in publications devoted to this process. Hydroquinone was added to the reaction mixture as an inhibitor of polymerization. Phosphorylated alkylacrylates and alkylmethacrylates were easily polymerized under the conditions of free radical catalysis at the elevated temperature [799].

Thermostable resins with a very low halogen content have been produced from monomers phosphorylated by phosphoric anhydride. They have been known for decades [800] and exhibit excellent flame-retardant properties [800]. Comparisons of the flame-retarding ability of the phosphorylated and the corresponding non-phosphorylated resins undoubtfully favoured the former substances. Therefore such materials are of pronounced importance for the construction industry particularly for reinforced articles including pipes [800].

Efficient self-extinguishing copolymers with unique optical properties were patented by Weil [801, 802] who applied the reaction of phosphoric anhydride (1 molar equivalent) with 3 mole equivalents of 2,3-dibromopropanol to produce monomers for homo- and copolymerization. Polymethyl metacrylate and copolymers of methylmatacrylate with unsaturated monomers are optically clear materials which have been successfully used as glass substitutes in the manufacture of lenses, windows, advertising displays, windshields, lighting fixtures, and so on [802]. The main disadvantage of such materials was their high flammability; attempts were made to minimize their flammability by introducing large quantities of tris(haloalkyl) phosphates and phosphonates. However these flame-retardant materials caused a loss of the high transparent properties. It was proposed [801, 802] to overcome this problem by using 2,3-dibromopropyl phosphoric acids and symmetrical bis(2,3-dibromopropyl) pyrophosphoric acid (Schemes 310, 311).

Methyl-2,3-dibromopropyl phosphoric acid was prepared by direct phosphorylation of the mixture of 2,3-dibromopropanol and methanol (Scheme 311). It was found to be a more efficient flame retardant as well as being less coloured. The ratio of methanol/2,3-dibromopropanol taken per one molar equivalent of phosphoric anhydride was $1:3$ [802]. Alternatively, the dialkylphosphoric acid could be synthesized by esterification of bis(2,3-dibromopropyl) pyrophosphoric acid by methanol.

$$Br-CH_2CHCH_2OH \quad \xrightarrow{P_4O_{10}} \quad Br-CH_2CHCH_2O-P(OH)_2 \quad +$$

(310)

$$+ \quad Br-CH_2CHCH_2O-P-O-P-OCH_2CHCH_2-Br$$

$$Br-CH_2CHCH_2O-P-O-P-OCH_2CHCH_2-Br \quad \xrightarrow{MeOH}$$

(311)

$$\longrightarrow \quad Br-CH_2CHCH_2O-P(OH)(OMe) \quad + \quad Br-CH_2CHCH_2O-P(OH)_2$$

The mixed product of such reaction could be used as a fire-retarding component without separation [802].

It was [802] admitted that there was no obvious explanation for the exceedingly high flame-retarding effect and low colouration of methyl 2,3-dibromopropyl phosphoric acid by methanol. The products of the reactions (Schemes 310, 311) were used as a fire-retardant alone or as a mixture of the two compounds. It was a very important observation [802] that close analogues such as bis(2,3-dibromopropyl) phosphoric acid and tris(2,3-dibromopropyl) phosphoric acid were found to have unsatisfactory fire-retarding activity in concentrations sufficient for the production of methyl methacrylate polymers with high optical clarity. Many types of methyl methacrylate copolymers that can be effectively fire retarded by dibromopropyl phosphoric acids are mentioned [802].

A flame-retarding agent was produced quantitatively in the reaction of phosphoric anhydride with tris(bromethyl) phosphite in boiling toluene [803]. Aaronson and Bright [803a] proposed oligomeric phosphate esters synthesized from epoxides and P_4O_{10} as efficient non toxic fire retardants.

4.3 EMULSIFIERS, LUBRICANTS AND ANTIWEAR ADDITIVES

Emulsifiers for food industry are prepared by heating triglycerides up to 30–100°C with a phosphoric anhydride–phosphoric acid mixture (70–100%) [818]. The salts of corresponding glycerophosphates have the required low toxicity.

The reaction of phosphoric anhydride with organosilicon compounds can produce silicon-containing polymer *organosilicons* [819] and *elastomers* [820–822]. The latter process (100–150°C) involves octamethylcyclotetrasiloxane.

The mixed zinc complex salts of phosphoric esters produced from phosphoric anhydride, hydroxy compounds and zinc salts have been patented as grease *antifriction additives* [804].

Well known for their lubricating, antirust and anticorrosion effects imidazoles were phosphorylated by phosphoric anhydride in the side chain of N-hydroxyalkylimidazolone (Alk=C_2H_4–$C_{32}H_{64}$) [805]. The overall composition of the mixtures formed was never analysed but one of the ingredients was said to be a corresponding monosubstituted phosphate (Scheme 312).

$$\text{(312)}$$

$$\text{Alk} = C_6H_{11} - C_{30}H_{61}; \qquad n = 1 - 6$$

$$\text{Solvent} = C_7H_{16}, \ Ph\,Me, \ C_6H_4Me_2$$

The mixtures of phosphorylated imidazoles exhibited *friction reducing* and *antiwear* properties which were more pronounced than their nonphosphorylated precursors in concentrations about 0.01 to about 10.0% by weight [805]. The most efficient concentrations were between 0.1 and 3.0% [805]. The products of phosphorylation were used in the forms of acidic phosphates and their ammonia salts. The further interaction of imidazolyl phosphates with carboxyamines, diamines, ether amines and ether diamines gave corresponding amidoesters of phosphoric acid [805] which also had useful antiwear properties.

Some other alkyl phosphates have been proposed as antirusting ingredients of propellant powder for gun burrels [806]. Trialkyl phosphates were proposed as corrosion protectors in compressors [807]. As anticorrosive agents alkyl (aryl) phosphates can be used neat as well as in mixtures with oils or various solvents [808].

Acidic alkylaryl phosphates, prepared by alkylation–phosphorylation [132] (Chapter 3, Section 1.2), and their salts are patented as useful *additives to lubricating oils.* Mono- and bis(alkylaryl) phosphoric acids can be utilized as *plasticizers* for polymeric materials including cellulose and its modified derivatives [809], phenolformaldehyde resins [810], including as ingredients of coating compositions and wetting agents [132].

The *metal cutting* the *antirusting* properties of monoalkyl, dialkyl phosphoric acids and phosphorylated alkylglycerols in combination with their high surface activity have led to their wide use for inhibiting corrosion [795, 796]. Alkyl phosphoric acids possess certain advantages over oil lubricants [789] due to their water solubility and hence easy removal of excess from the surface of metal parts.

4.4 CRYSTAL GROWTH MODIFIERS

The *photographic materials industry* exploits the pronounced *antistatic* action of dialkyl phosphates in the outer gelatin layer of photographic films [822a]. For example methyldodecyl phosphate was introduced in gelatin in the form of a methanol solution.

Zavlin and Efremov [354, 356, 357, 823–827] used some organic compounds phosphorylated by phosphoric anhydride for *modifying silver halide microcrystals* in photosensitive films and papers. The technological process of silver halide microcrystals modifying [828–830] in photographic materials resulted in 6–10 times higher sensitivity of photographic emulsions [831–835]. The crystal modifying activity of fluoroglucinols phosphorylated by phosphoric anhydride (Chapter 3, Section 8.2), azoles (Chapter 3, Section 3.5), pyrimidinetriones (Chapter 3, Section 3.4) and some other compounds was studied [163, 354, 823–827]. Black-and-white and colour silver halide (AgBr, AgCl/AgBr and AgI/AgBr) models as well as industrial emulsions were included. Phosphorylated azoles effectively transformed the cubic silver halide microcrystals into rombododecahedral crystalline structures. According to electronic microscopy different adsorbants modified silver halide cubic (Figure 6.1a) emulsions into rombododecahedral (Figure 6.1b) or/and ball-shaped microcrystals.

The less active additives–modifiers of silver halide emulsions produced crystals of the intermediate ball-shape structure.

The shape of the microcrystals depended not only upon the nature of modifiers but also their concentration. The optimum concentration of the most active modifier in silver halide emulsions was found to be $c.\,200\,mg/Ag$ g-atom [823–825]. Higher concentrations depressed the emulsions' sensitivity (S) probably due to shielding by the absorbant on the modified microcrystals' surface. Other parameters for the efficient modification were shown to include temperature, pAg, pHal and some others [823–825]. The specific features of emulsions modified by the phosphorylated azoles emulsions were not only high sensitivity but antifog (D_0) and a highly pronounced stabilizing effect along with good contrast (γ) data (Table 6.1).

Table 6.1 Photographic data of the control and modified AgBr model emulsions

C, $mg_{modifier}$/Ag g-atom	S	γ	D_0
0	12.0	4.5	0.12
100	25.0	5.0	0.03
200	37.0	4.9	0.02
300	29.0	4.7	0.05

For example the stabilizing effect of benzotriazolyl phosphoric acid was much higher than that of widely used stabilizer benzotriazole — the precursor of benzotriazolido phosphate.

The technological experiments with phosphorylated silver halide crystal growth modifiers were accompanied by a study of their adsorption on the crystals as an approach to the actual mechanism of modification. The isotherms of adsorption of phosphorylated modifiers and their nonphosphorylated precursors contained distinctive portions of saturation. For the former adsorbates the concentrations of saturation were lower than for the latter ones which means that the phosphoryl group promoted adsorption of modifiers by the additional interaction between the electrophilic phosphorus group and subsurface silver cations. At pH about 9 in the range of concentrations under study (100–400 mg/Ag g-atom) the adsorption was determined to have a reversible character. The most efficient adsorption on AgBr microcrystals was shown to be for benzazoles phosphorylated by phosphoric anhydride [823–825].

It was established [836], by means of solid state ^{13}C NMR spectrometry, that the aromatic rings were of insignificant influence upon the process of adsorption of organic compounds on the surface of silver halide microcrystals whereas the ^{31}P NMR spectra of the adsorbed and nonadsorbed species varied markedly which suggested that the phosphorus groups were the crucial parts of the modifiers in the adsorption process. This effect explains well the lower modifying activity of nonphosphorylated compounds compared to their phosphorylated derivatives.

Phosphorylated pyrimidinetriones exhibited a distinctive modifying activity in silver halide photographic emulsions with cubic and tabular microcrystals.

On the basis of the experimental data Efremov and Zavlin [354, 823–825] worked out the main principles for target orientated synthesis of active silver halide crystal growth modifiers. These are:

1. Silver halide crystal growth modifiers need to be cyclic or compact acyclic compounds containing three or more n-electron pair centres for an efficient interaction with the subsurface silver cations.
2. The crystal growth modifiers do not need to be aromatic compounds. The best adsorbates towards the silver halide crystals' surface are those with

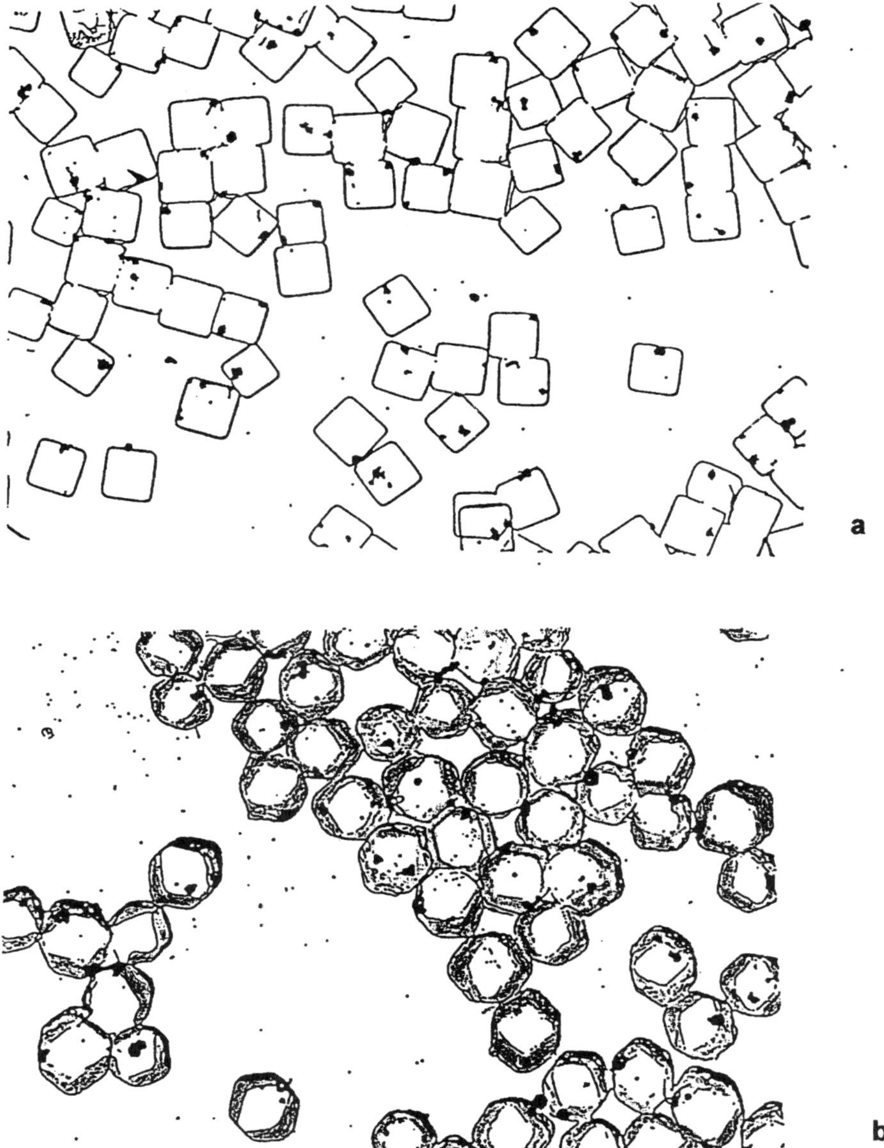

Figure 6.1 Electronographs of the core cubic AgBr emulsion (a) and modified rombododecahedral AgBr emulsion (b) [Authors]

distinctive steric lability, in other words compounds that can easily adapt to the stereochemistry of the crystals' surface.

3. Tautomeric (prototropic and/or phosphorotropic) equilibria in the adsorbates' molecules promote the modifying activity towards micro-crystals.

4. Introduction of dihydroxyphosphoryl or di(tri)hydroxypyrophosphoric fragments in the molecules of modifiers enriches their interaction with a) the subsurface silver cations on one hand and b) electrophilic phosphorus atoms with surface halogens anions on the other.

4.5 INSECTICIDES, FUNGICIDES, EXTRACTANTS AND OTHER APPLICATIONS

Insecticides and fungicides Many years after Dupuis's publication [88] on phosphorylation of phenols by phosphoric anhydride (Chapter 3, Section 1.2) Allen and Parson's patented [89] nematosides which were synthesized by the same method and efficient against root-knot on tomato plants (Scheme 313). For better water solubility diphenylphosphoric acids were produced in the form of their diethylammonium salts.

$$\left(313\right)$$

Insecticides were prepared from alkylphosphonates and phosphoric anhydride at 50–70°C (1–2 h) [837]. The best results were achieved for the lower homologues (C_{1-4}) and a molar ratio $P_4O_{10}:RP(O)(OR')_2=1:2-12$.

The manufacture of the efficient insecticide TEPP is based on the reaction of phosphoric anhydride with triethyl phosphate [638] (Scheme 314).

$$P_4O_{10} + 8(RO)_3P = O \longrightarrow 6(RO)_2P(O) - O - P(O)(OR)_2 \qquad (314)$$

The mixture of phosphoryl chloride with phosphoric anhydride was used as a chlorinating agent in the synthesis of 2- and 4-chloro-1,5-naphthyridenes [838]. 1,5-Naphthyridine was oxidized by peracetic acid into a mixture of 1,5-naphthyridine mono- and di-N-oxides (Scheme 315). The former product was isolated and heated in phosphoryl chloride with phosphoric anhydride for

30 min to form the corresponding chlorinated compounds the ratio of which was determined by gas chromatography [838] (Scheme 315).

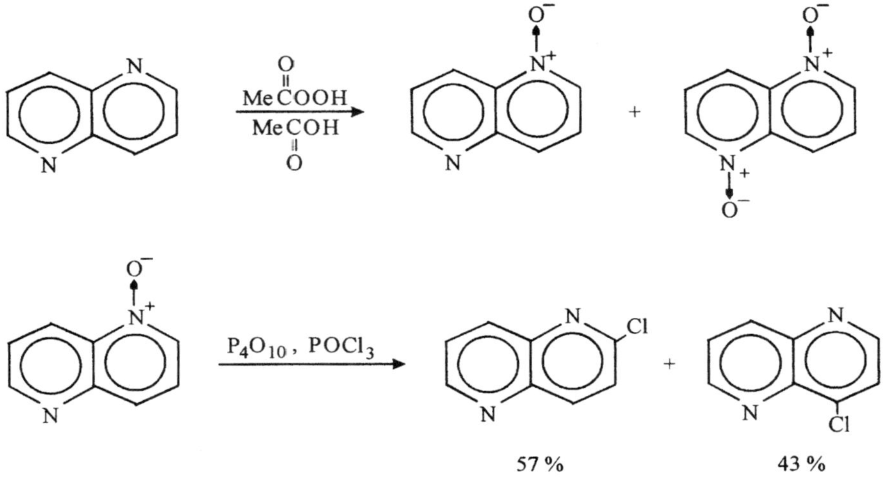

$$(315)$$

Mixed alkylphenyl ethers which are also used as insecticide additives, were synthesized from alkyl (ethyl, butyl, sec-butyl, amyl) esters of phosphoric acid and phenol [839].

Extractants Alkyl phosphoric acids derived from phosphoric anhydride and alkanols formed *complexes with uranium and vanadium* [840], which favoured the extraction of heavy metals. The extraction could be performed in kerosene at pH about 1.0 for uranium and at pH 2.0–3.0 for removing vanadium. The recovery rate of uranium achieved was as high as 95–99%. Though amines are said to be more selective for uranium than alkyl phosphoric acids, the latter are aproximately four times cheaper than amines [840].

The most active ligand for uranium extraction from wet-process phosphoric acid was reported to be [841] decyl ester of pyrophosphoric acid. The extraction was carried out in kerosene followed by precipitation of the uranium salt by hydrofluoric acid. The technological scheme for uranium extraction has been described [841] which includes the synthesis of decyl pyrophosphoric acid.

Another technological process proposed for the extraction of uranium utilised di(2-ethylhexyl) phosphoric acid produced from phosphoric anhydride and the corresponding alcohol (Chapter 3, Section 1.2) mixed with tributyl phosphate in high flash-point (64°C) kerosene [842]. The proposed mixture was not only an efficient process but the most favorable from an economical point of view.

Bis(2-ethylhexyl) phosphoric acid has also been used for binding cadmium in the corresponding cadmium salt (Scheme 316).

$$Cd^{2+} + (CH_3CH_2CH_2CH_2CHCHO)_2POH$$
$$\underset{Et}{\mid} \quad \underset{O}{\parallel}$$
$$\longrightarrow [(CH_3CH_2CH_2CH_2CHCHO)_2PO]_2Cd + 2H^+ \tag{316}$$
$$\underset{Et}{\mid} \quad \underset{O}{\parallel}$$

Other applications 2-Ethyl-1,3-hexanediol is phosphorylated by phosphoric anhydride [767] and the resulting alkylphosphate is used to modify the epoxyresins to assist the insertion of hydroxylamine in aminoplast coating.
Antistatic action is another valuable property of organic phosphates derived from phosphoric anhydride. They are widely used in the textile industry especially in the processes of spinning cotton and wool [774, 775] where the tendency to build up electrostatic charge is particularly pronounced.
Finishing agents for textiles, and *antistatic agents* for plastics, are synthesized from P_4O_{10} and polyethyleneglycols [768]. However, Romanian chemists [769] consider $POCl_3$ to be a better phosphorylating agent for polyethyleneglycol than phosphoric anhydride.

In concentrations of about 2% alkylammonium phosphates are important ingredients of *antifreeze* compositions.

The reaction of phosphoric anhydride with diethylether is of pronounced practical importance as it is used for the industrial production of triethyl phosphate. The process is carried out at 180°C in an atmosphere of ethylene (Scheme 317).

$$P_4O_{10} + 6(C_2H_5)_2O \longrightarrow 4(C_2H_5O)_3P=O \tag{317}$$

Perfluoroalkyl nitriles prepared by the dehydration of the corresponding acid amides by phosphoric anhydride (Chapter 3, Section 3.8.1) were converted into perfluoronitrile polymers [811] that possess *elastomeric* properties, thermostability and chemical resistance [811–813].

The role of phosphoric anhydride in the production of *catalysts* is vast. The petroleum industry has made extensive use of its ability to improve selectivity and conversion efficiency. Phosphoric anhydride is used widely in industry as a catalyst ingredient or as a promoter for the vapour-phase catalytic syntheses of maleic anhydride, acrolein, acrylic and methacrylic acids, aldehydes, nitriles, butadiene and others.

Recently the reaction of phosphoric anhydride with epoxy compounds was used for synthetic polymer modification [814, 815] and for the synthesis of polymers from cyclohexene oxide [816] and epoxychlorohydrin in the presence of aniline [817]. The latter method produced chlorohydrin *rubbers* with improved processability.

4.6 HANDLING, HEALTH AND SAFETY

Several factors cause considerable difficulties in industrial applications of phosphoric anhydride. Most of the problems arise from its extraordinary hydrophilicity which affects the storage and methods of the addition of phosphoric anhydride to reaction vessels. It must be stored in air- and moisture-tight drums that preferably should be emptied in one go because of the danger of hydrolysis of P_4O_{10} in a partially filled drum even with all the necessary precautions taken. Even small amounts of phosphoric anhydride powder on a rubber gasket can cause significant contamination.

Feeding a reactor with phosphoric anhydride must therefore be performed under strictly moisture-free conditions with intensive (preferably continuously) agitation to distribute phosphoric anhydride evenly in the reaction mixture without being thrown onto the walls of a reactor above the liquid level and avoiding the formation of lumps. The latter can initiate local different and often undesirable processes from the required reaction. The chemical companies dealing with phosphorylation by phosphoric anhydride have designed various feed hoppers and rotary valves constructed of stainless steel or alternatively used continuous purging with an inert dry gas.

Phosphoric anhydride is considered to be a corrosive material and therefore it must be handled with precaution using adequate personal protection equipment.

According to Samuel Moore, Vice-President (Research and Development) of Burlington Chemical Co (North Carolina, USA), the future of phosphates, phosphorylated polymers and phosphorylated intermediates is a fruitful area of product development for the company, and future developments appear bright.

References

[1] Corbridge D.E. *Phosphorus*. Elsevier, Amsterdam, 1990, pp. 69–74, 93–110.
[2] Kariakin Yu.V. and Angelov I.I. *Pure Chemical Compounds*. "Khimia", Moscow, 1974, p. 63.
[3] Van Wazer, J.R. *Phosphorus and its Compounds*, Russian edition, **1**, Inostr. Liter., Moscow, 1962.
[4] Glixelli S. and Boratynski K. Z. *Anorg. Allgem. Chem.* (1938) **235** (3), 225–41.
[5] Muenow D.W., Uy O.M. and Margrave J.L. *J. Inorg. Nucl. Chem.* (1970) **32** (11), 3459–67.
[6] Andrews L. and Withnall R. *J. Am. Chem. Soc.* (1988) **110** (17), 5606–11.
[7] Mielke Z., Andrews L. *J. Phys. Chem.* (1989) **93**, 2971–6.
[8] Glonek T., Van Wazer J.R., Kleps R.A. and Myers T.C. *Inorg. Chem.* (1974) **13** (10), 2337–45.
Glonek T., Van Wazer J.R., Mudgett M. and Myers T.C. *Inorg. Chem.* (1972) **11** (3), 567–70.
[9] Glonec T., Myers T.C., Han P.Z. and Van Wazer J.R. *J. Am. Chem. Soc.* (1970) **92**, 7214–7.
[10] Glonek T., Myers T.C. and Van Wazer J.R. *J. Am. Chem. Soc.* (1975) **97** (1), 206–7.
[11] Kreshkov A.P., Drozdov V.A. and Orlova I.Yu. Zh. *Obsch. Khim.* (1966) **36** (3), 525.
[12] Hoeflake J.M. and Scheffer F.E. *Rec. Trav. Chim.* (1926) **45**, 191–206.
[13] Smits A., Ketelaar J.A.A. and Meyering J.L. *Z. Phys. Chem. (B)* (1938) **65**, 87–97.
[14] de Decker H.C.J. *Rec. Trav. Chim.* (1941) **60** (5), 413–27.
[15] de Decker H.C.J. and MacGillavry C.H. *Rec. Trav. Chim.* (1941) **60** (3), 153–75.
[16] Hill W.L., Faust G.T. and Hendricks S.B. *J. Am. Chem. Soc.* (1943) **65** (2), 794–802.
[17] MacGillavry C.H., de Decker H.C.J. and Nijland L.M. *Nature* (1949) **164** (10), 448–9.
[18] Cruickshank D.W.J. *Acta Cryst.* (1964) **17** (Part 5), 679–80.
[19] Jansen M. and Luer B. *Z. Kristallogr.* (1986) No. 177. 149–51.
[20] Tilden W.A. and Barnett R.E. *J. Chem. Soc.* (1896) 154–60.
[21] Mayer C. *Berichte* (1879) **12**, 610.
[22] Khodakov Yu.V. *Dokl. Akad. Nauk USSR.* (1944) **43** (5), 212–5.
[23] Maxwell L.R., Hendricks S.B. and Deming L.S. *J. Chem. Phys.* (1937) **5**, 626–37.
[24] Hampson G.C. and Stosick A.J. *J. Am. Chem. Soc.* (1938) **60** (8), 1814–22.
[25] Akishin P.A., Rambidi N.G. and Zasorin E.Z. *Kristallogr.* (1959) **4** (3), 360–4.
[26] Cruickshank D.W.J. *Acta Cryst.* (1964) **14** (Part 5), 671–2.

[27] Cruickshank D.W.J. *Acta Cryst.* (1964) **17** (Part 5), 672–3.

[28] Cruickshank D.W.J. *Acta Cryst.* (1964) **17** (Part 5), 674–5.

[29] Cruickshank D.W.J. *Acta Cyst.* (1964) **17** (Part 5), 675–6.

[30] Cruickshank D.W.J. *Acta Cyst.* (1964) **17** (Part 5) 677–9.

[31] Cruickshank D.W.J. *J. Chem. Soc.* (1961) 5486–8.

[32] Beagley B., Cruickshank D.W.J., Hewitt T.G. and Haaland A. *Trans. Faraday Soc.* (1967) **63** (4) (532), 836–45.

[33] Beagley B., Cruickshank D.W.J., Hewitt T.G. and Jost K.H.. *Trans. Faraday Soc.* (1969) **65** (5) (557), 1219–30.

[34] Jost K.H. and Schneider M. *Acta Cryst. B.* (1981) **37**, 222–4.

[35] Jansen M. and Mobs M. *Inorg. Chem.* (1984) **23** (26), 4486–8.

[36] Stachel D., Svoboda I. and Fuess H. *Acta Crystallogr. C. Cryst. Struct. Commun.* (1995) **C 51** (6), 1049–50.

[37] Lohr, Jr. L.L. *J. Phys. Chem.* (1990) **94** (5), 1807–11.

[38] Slivko S.A. and Krivoviazov E.L. *Zh. Neorg. Khim.* (1993) **38** (11), 1860–5.

[39] Heinz D.Z. *Anorg. Allg. Chem.* (1965) **336** (3–4), 137–55.

[40] Gillespie R.J. and Robinson E.A. *Canad. J. Chem.* (1964) **42** (11), 2496–503.

[41] Jost K.H. *Acta Cryst.* (1964) **17** (12), 1593–8.

[42] Cyvin S.J. and Cyvin B.N. *Z. Naturforsch.* (1971) **262** (5), 901–6.

[43] Berzelius J.J. *Lieb. Ann.* (1843) **46**, 251–81.

[44] Corbridge D.E.C. *The Structural Chemistry of Phosphorus Compounds. Topics in Phosphorus Chemistry*, 1966, Vol. 3, p. 57.

[45] Hoffman H. and Becke-Goehring M. *Phosphorus Sulfides. Topics in Phosphorus Chemistry*, 1976, Vol. 8, p. 193.

[46] Gerding H. and de Decker H.C.J. *Rec. Trav. Chim.* (1946) **64**, 191–3.

[47] Sidorov T.A. and Sobolev N.N. *Optika Spektrosc.* (1957) **2** (6), 717–27.

[48] Vinter J.G., Davis A. and Saunders M.R.J. *Computer-Aided Mol. Des.* (1987) No. 1, 31–51.

[49] Vinter J.G. "Molecular modelling systems" in *Molecular Graphics and Drug Design*, ed. A.S.V. Burgen, G.C.K. Roberts, M.S. Elsevier. Amsterdam, New York, Oxford, 1986, pp. 16–56.

[50] Dewar M.J.S., Zoebisch E.G., Healy E.F. and Stewart J.J.R. *J. Am. Chem. Soc.* (1985) **107** (13), 3902–9.

[51] Stewart J.J.P. *J. Comp. Chem.* (1989) **10** (2), 221–64.

[52] Stewart J.J.P. *J. Comp. Chem.* (1989) **10** (2), 209–20.

[53] Shimazaki K., Mori M., Okada K., Chuman T., Goto H., Osawa E., Sakakibara K. and Hirota M. *J. Chem. Soc. Perkin Trans. 2* (1992) **(3)** 811–8.

[54] Jin A.Y. and Benesch L.A. *Can. J. Chem.* (1994) **72** (6), 1596–604.

[55] Andrews E., Katritzky A.R., Malhotra N. and Stevens J. *J. Org. Chem.* (1992) **57** (13), 3698–705.

[56] Pauling L. *General Chemistry*, Dover Publications, New York, 1988, p. 278.

[57] Glushko V.P. *et al. Thermodynamic Properties of Individual Compounds*, Nauka, Moscow, 1978, Vol. 1, Part 2, p. 280.

[58] *Thermochemical Properties of Inorganic Substances*, Springer, Berlin, 1991. Vol. 2, p. 1527.

[59] Jost D.M. and Russell H., Jr. *Systematic Inorganic Chemistry*, Prentice-Hall, Inc, New York, 1944, p. 179–82.

[60] *Phosphorus Pentoxide*, Stauffer Chemical Company, Technical Report, Connecticut, 1978.

[61] Rabinovich V.A. and Khavin Z.Ia. *Chemical Reference Book*. Khimia, Leningrad, 1977, p. 108.

[62] Gerding H. and Van Brederode H. *Rec. Trav. Chim.* (1946) **64**, 183–190.

[63] Daasch L.W. and Smith D.S. *Anal. Chem.* (1951) **23** (6), 853–68.

[64] Beattie I.R., Livington K.M.S., Ozin G.A. and Reynolds D.J. *J. Chem. Soc. A.* (1970) (**3**) 449–51.

[65] Robinson E.A. *Canad J. Chem.* (1963) **41** (12), 3021–33.

[66] Withnall R. and Andrews L. *J. Phys. Chem.* (1988) **92** (16), 4610–19.

[67] Lazarev A.N., Mirgorodsky A.P. and Ignatov I.S. *Vibrational Spectra of Complex Oxides*. Nauka, Leningrad, 1975, pp. 248–9.

[68] Chapman A.C. *Spectrochim. Acta. A.* (1968) **24**, 1687–96.

[69] Slivko S.A. and Krivovyazov E.L. *Zh. Neorg. Khim.* (1993) **38** (10), 1606–10.

[71] Kosolapoff G.M. *Organophosphorus Compounds*, Wiley, New York, 1950.

[72] Rauhut M.M. "Synthesis of Organophosphorus Compounds from Elemental Phosphorus" in *Topics in Phosphorus Chemistry*, ed. Greyson M. and Griffith E.J. Wiley, Interscience Publishers, New York, 1964.

[73] Glixelli S. *Ann. Soc. Chim. Polonorum.* (1940) **18**, 515–24.

[74] Glixelli S. and Boratinsky K. *Z. Anorg. Allg. Chem.* (1940) **235**, 224–32.

[75] Rodionova N.I. and Khodakov Yu.V. *Zh. Obsch. Khim.* (1950) **20** (8), 1347–57.

[76] Burkhardt G., Klein M.P. and Calvin M. *J. Am. Chem. Soc.* (1965) **83** (3), 591–6.

[77] Bell R.N., Andrieth L.F. and Hill O.F. *Ind. Eng. Chem.* (1952) **44** (3), 568–72.

[78] Thilo E. and Wieker W. *Z. Anorg. Chem.* (1954) **277**, 27–36.

[79] Ratz R. and Thilo E. *Ann. Chem.* (1951) **572** (3), 173–89.

[80] Kalliney S.Y. "Cyclophosphates" in *Topics in Phosphorus Chemistry.* Vol. 7, ed. Griffith E.J. and Grayson M. Wiley, Interscience, New York, 1972, pp. 268–9.

[81] Seel F., Ballreich K. and Schmutzler R. *Ber.* (1961) **94**, 1173–84.

[82] Gerber A.B. and Miles F.T. *Industr. Eng. Chem. Analyt. Edition* (1941) **13** (6), 406–12.

[83] Henry W.M., Nickless G. and Pollard F.H. *J. Inorg. Nucl. Chem.* (1967) **29**, 2479–80.

[84] Shima M., Hamamoto K. and Utsumi S. *Bull. Soc. Chim. Japan* (1960) **33** (10), 1386–91.

[85] Dillon K.B. and Waddington T.C. *J. Chem. Soc. (A)* (1970) 1146–50.

[86] Osterheld R.K. "Nonenzymatic Hydrolysis" in *Topics in Phosphorus Chemistry*, Vol. 7, ed. Griffith E.J. and Grayson M. Wiley, Interscience, New York, 1972, pp. 181–9.

[87] Schwarz F.Z. *Anorg. Chem.* (1895) **260**, 249–66.

[88] Dupuis M.P. *Bull. Soc. Chim. France* (1910) **7**, 846–7.

[89] Allen G.J. and Parsons R.C. US Pat. 2,903,393 (1960).

[90] Grun A. and Limpacher R. *Ber.* (1926) (5), 1345–50.

[91] Grun A. and Limpacher R. *Ber.* (1926) (5), 1350–60.

[92] Grun A. and Limpacher R. *Ber.* (1927) (1), 147–50.

[93] Grun A. and Limpacher R. *Ber.* (1927) (1), 151–6.

[94] Jungermann E. and Silberman H.C. "Phosphorus-containing Anionic Surfactants" in *Anionic Surfactants*, Part 2, ed. W.M. Linfield, Surfactant Science Series, Vol. 7, Dekker, New York, 1976, pp. 504–14.

[95] Plimmer R.H.A. and Burch W.J.N. *J. Chem. Soc.* (1929) 279–91.

[96] Cherbuliez E., Colak-Antic S., Schwarz M. and Rabinowitz J. *Helv. Chim. Acta.* (1963) **46** (161), 1448–52.

[97] Calwell J.R. US Pat. 3,153,080 (1964).

[98] Cherbuliez E., Leber J.-P. and Schwarz M. *Helv. Chim. Acta.* (1953) **36** (161), 1189–99.

[99] Cherbuliez E. and Weniger H. *Helv. Chim. Acta.* (1945) **28**, 1584–91.

[100] Ivanov B.E., Zheltukhin V.F. and Eliseenkov V.N. *Chemistry and Applic. Organophosph. Comp. VIII Vsesouzn. Conf. (USSR)* Nauka, Moscow, 1987, pp. 259–64.

[101] Petrov K.A. and Treschalina L.V. *Itogi Nauki i Tekhniki. Technol. Organ. Veschestv.* (USSR) (1977) **4**, 60–94.

[102] Okamoto Y. *Bull. Chem. Soc. Japan.* (1985) **58** (11), 3393–4.

[103] Cotton F.A. and Wilkinson G. *Advanced Inorganic Chemistry*, Wiley, New York, 1966, p. 499.

[104] Vila Busquets P. Pat. 421396, Spain (1977).

[105] Antony R. and Pillai C.K.S. *J. Appl. Polym. Sci.* (1993) **49**, 2129–35.

[106] Smith H.G. and Cantrell T.L. US Pat. 2,441,295 (1948).

[107] Adler H. and Woodstock W.H. *Chem. Ind.* (1942) **51**, 516.

[108] Elias A., Azzouz A. and Rodehuser L. *Phosphorus Sulfur.* (1993) **85** (Part 1–4), 91–9.

[108a] Perka J., Ropuszinski S. and Mularczyk E. *Przem. Chem.* (1978) **57** (2), 64–6. Ignatenko A.V., Pastukhova I.V., Petrov K.A., Smelov V.S., Chauzov V.A. and Schmidt V.S. *Zh. Anal. Khim.* (1982) **37** (4), 744–7.

[109] Tebby J.C. *Handbook of Phosphorus—31 Nuclear Magnetic Resonance Data*, CRC Press, Boca-Raton, 1991, pp. 262–5.

[110] Iasnikov A.A., Markovski L.N., Galichev V.V., Kozlova N.Ia., Smetanin V.V., Lozinski M.O., Liptuga N.I., Malovik V.V. and Zavadski V.N. Pat. 1033501. USSR (1981).

[111] US Pat. 1,962,827 (1934).

[112] Clermont L.P. US Pat. 3,565,886 (1971).

[113] Clermont L.P. Pat. 868,344. Canada (1971).

[114] Benning A.F. US Pat. 2,597,702 (1952).

[115] Joyce R.M. US Pat. 2,559,628 (1951).

[116] Hays H.H. and Kosolapoff G.M. US Pat. 3,444,522 (1969).

[117] Astr. Pat. 310,907 (1973).

[118] T. Hasigawa US Pat. 3,686,371 (1972).

[119] Milwidsky B.M. and Gabriel D.M. *Detergent Analysis. A Handbook for Cost-effective Quality Control*, George Godwin, London, 1982, p. 137.

[120] Pat. 07,330,784. Japan (1995).

[121] Pat. 07,316,170. Japan (1995).

[122] O'Lenick, A.J. Jr. and Parkinson J.K. *Textile Chemist & Colorist* (1995) **27** (11), 17–20.

[123] Nusslein J. *Parfum. Cosmet. Savous.* (1959) **2**, 554–62.

[124] Woodstock W.H. US Pat. 2,586,897 (1952).

[125] Kurosaki T., Wakatsuki J., Imamura T., Matsunaga A., Furugaki H. and Sassa Y. *J. Com. Esp. Deterg.* (1988) **19** (1), 191–205.

[126] Cherbuliez E. *Helv. Chim. Acta.* (1963) **46**, 2464.

[127] Tsubone K., Uchida N., Niwase H. and Honda K. *J. Am. Oil Chemist's Soc.* (1989) **66** (6), 829–33.

[128] Sedlacek J., Ranny M. and Kondelik P. *Chem. Prumysl.* (1990) **40/65** (6), 283–7.

[129] Prahl W.H. Pat. 763,311. Brit. (1956).

[130] Yamagata Y., Kojima H., Ejiri K. and Inomata K. *Origins of Life* (1982) **12** (4), 333–7.

[131] Ganiev R.G. Pat. 1,514,748 USSR (1989).

[132] Otto F.P. US Pat. 2,745,863 (1956).

[133] Clarke F.B. and Lyons J.W. *J. Am. Chem. Soc.* (1966) **88** (19), 4401–5.

[134] Nielsen M.L., Pustinger J.V. and Strobel J. *J. Chem. Eng. Data* (1964) **9**, 167.

[135] Cherbuliez E. and Wenige H. *Helv. Chim. Acta.* (1946) **29**, 2006.
[136] Woodstock W.H. US Pat. 2,568,784 (1951).
[137] Malatesta L. and Laverone F. *Gazz* (1951) **81**, 596–608.
[138] Langheld K. *Ber.* (1910) **43** (10), 1857–60.
[139] Langheld K. *Ber.* (1911) **44**, 2076–87.
[140] Pascal M.P. *Acad. Sci. Compt. Rend.* (1923) 176, 1398–1400.
[141] Wertyporoch E. and Kiekenberg H. *Biochem. Z.* (1934) **268** (1–3), 8–16.
[142] Steinkopf W. and Schubart I. *Ann.* (1921) **424** (1), 1–23.
[143] Thilo E. and Woggon H. *Z. Anorg. Chem.* (1954) **277** (1–2), 17–26.
[144] Van Wazer J.R., Callis C.F., Shoolery J.N. and Jones R.C. *J. Am. Chem. Soc.* (1956) **78** (22), 5715–26.
[145] Weill G., Klein M. and Calvin M. *Nature* (1963) **200** (4910), 1005–6.
[146] Pollmann W. and Schramm G. *Biochim. Biophys. Acta.* (1964) **80** (1), 1–7.
[147] Parks J.R. and Van Wazer J.R. *J. Am. Chem. Soc.* (1957) **79** (18), 4890–7.
[148] Cramer F. and Hettler H. *Chem. Ber.* (1937) **70**, 1458–69.
[149] Adler H. US Pat. 2,462,057 (1949).
[150] Hull D.C. and Snodgrass J.R. US Pat. 2,407,279 (1946).
[151] Hull D.C. and Snodgrass J.R. US Pat. 2,508,389 (1950).
[152] Hull D.C. and Snodgrass J.R. US Pat. 2,492,153 (1949).
[153] Kochetkov N.K., Budowsky E.L., Domkin V.D. and Khromov-Borissov N.N. *Biochim. Biophis. Acta.* (1964) **80** (1), 145–8.
[154] Jacob T.M. and Khorana H.G. *J. Am. Chem. Soc.* (1964) **86** (8), 1630–5.
[155] Schwarzmann E. and Van Wazer J.R. *J. Am. Chem. Soc.* (1961) **83** (1), 365–7.
[156] Van Wazer J.R. and Norval S. *J. Am. Chem. Soc.* (1966) **88** (19), 4415–23.
[157] Schramm G. and Berger H. *Z. Naturforsch.* (1967) **22b** (6), 587–96.
[158] Shima M., Hamamoto K. and Utsumi S. *Bull. Soc. Scim. Japan.* (1960) **33** (10), 1386–91.
[158a] Wilcox R.D., Harris G.H. and Olson R.S. *Tetrahedron Lett.* (1968) **57**, 6001–2.
[159] Lottermoser M. Pat. 3,400,055 Ger. (1985).
[160] Staudinger H. and Rathsam G. *Helv. Chim. Acta.* (1922) **5**, 645–55.
[161] Brannock K.C. *J. Am. Chem. Soc.* (1951) **73** (10), 4953–4.
[162] Diels O. and Wolf B. *Ber.* (1906) **39**, 689–97.
[163] Efremov D.A., Zavlin P.M., Essentseva N.S. and Tebby J.C. *J. Chem. Soc. Perkin Trans. 1.* (1994) 3163–8.
[164] Zavlin P.M., Efremov D.A., Essentseva N.S. and Tebby J.C. *Zh. Obsch. Khim.* (1995) **65** (4), 591–4.
[165] Zavlin P.M. and Efremov D.A. *Phosphorus Sulfur.* (1990) **49/50**, 247–50.
[166] Becke-Goering M. and Sambeth J. *Z. Anorg. Chem.* (1958), **297**, 287–95.
[167] Moeller T. *Inorganic Chemistry*, Wiley, New York, 1952, pp. 658–60.
[168] Jones O.C. and Arvan P.G. US Pat. 2,717,198 (1955).
[169] Fisher H. *Uber Reactionsproducte aus Phosphorpentoxyd und Ammoniak und deren Einung zur Wasserenthartnung*, M. Dittert, Dresden, 1941, p. 34.
[170] Sanfourche A., Hernette A. and Fau M. *Bull. Soc. Chim. France* (1930) **47**, 273–9.
[171] Harris L. and Woodster C.B. *J. Am. Chem. Soc.* (1929) **51**, 2121–6.
[172] Woodstock W.H. US Pat. 2,122,122 (1938).
[173] Sancho J. and Moles E. *Annales Soc. Espan. Fis. Quim.* (1932) **30**, 701–6.
[174] Fukumura C., Inine K., Iwata M. and Navita N. Eur Patt. Appl. EP 494,778. *Chem. Abstr.* (1993) **118**, 4183.
[175] Staffel T., Becker W. and Neumann H. Ger. Offen. DE 4133811. *Chem. Abstr.* (1993) **118**, 237–73.

[176] Buck A.C. and Lankelma H.P. *J. Am. Chem. Soc.* (1948) **70** (7), 2398–400.

[177] Wolf G.-U. and Meisel M. *Z. Chim.* (1982) **1**, 54.

[178] Behrens H. and Kinzel K. *Z. Anorg. Chem.* (1959) **299**, 241–51.

[179] Buck A.C., Bartleson J.D. and Lankelma H.P. *J. Am. Chem. Soc.* (1948) **70**, 744–6.

[180] Jander G. and Weis J. *Z. Electrochem.* (1957) **61**, 1275–83.

[181] Buck A.G. and Lankelma H.P. *J. Am. Chem. Soc.* (1948) **70**, 744–6.

[182] Wise G. and Lankelma H.P. *J. Am. Chem. Soc.* (1952) **74**, 529–31.

[183] Griffith E.J., Ponnamperuma C. and Gabel N.W. *Origins of Life* (1977) **7** (8), 71–85.

[184] Kametani T. *Heterocycles.* (1981) **16** (7), 1205–42.

[185] Nifantyev E.E. and Gratchev M.K. *Uspekhi Khim.* (1994) **63** (7), 602–37.

[186] Zabirov N.G., Shamsevaleev F.M. and Cherkasov R.A. *Uspekhi Khim.* (1991) **60** (10), 2189–219.

[187] Kitano M. and Kuchitsu K. *Bull. Chem. Soc. Japan.* (1974) **47** (3), 631–4.

[188] La Planche L.A. and Roger M.T. *J. Am. Chem. Soc.* (1964) **86** (3), 337–41.

[189] Stewart W.E. and Siddall T.H. *Chem. Rev.* (1970) **70** (5), 517–51.

[190] Yoder C.H., Sandberg J.A. and Moore W.S. *J. Am. Chem. Soc.* (1974) **96** (7), 2260–2.

[191] Challis B.C. and Challis J.A. *Amides and Related Compounds. Comprehensive Organic Chemistry. The Synthesis and Reactions of Organic Compounds. Vol. 2*, Pergamon Press, Oxford, 1979, pp. 957–1065.

[192] Zavlin P.M., Efremov D.A. and Tebby J.C. *Zh. Obsch. Khim.* (1994) **64** (8), 1308–9.

[193] Grout R.J. *Biological Reactions and Pharmaceutical Uses of Imidic Acid Derivatives. The chemistry of amidines and imidates*, ed. S. Patai, Wiley, London, 1975, p. 257.

[194] Mizobuchi K. and Buchanan J.M. *J. Biol. Chem.* (1968) **243**, 4853–62.

[195] Mizobuchi K., Kenyon G.L. and Buchanan J.M. *J. Biol. Chem.* (1968) **243**, 4863–77.

[196] Zimin M.G., Afanasiev M.M. and Pudovik A.N. *Zh. Obsch. Khim.* (1979) **49** (11), 2621–2.

[197] Zabirov N.G., Shamsevaleev F.M., Scherbakova V.A., Cherkasov R.A., Soloviev V.N., Chekhlov A.N., Dmitrieva G.V. and Martinov I.V. *Zh. Obsch. Khim.* (1990) **60** (9), 1997–2004.

[198] Iarkova E.G., Safiullina N.P., Chistiakova I.G., Zabirov N.G., Shamsevaleev F.M., Scherbakova V.A. and Cherkasov R.A. *Zh. Obsch. Khim.* (1990) **60** (9), 2005–11.

[199] Zabirov N.G., Soloviev V.N., Shamsevaleev F.M., Cherkasov R.A., Chekhlov A.N., Tsifarkin A.G. and Martinov I.V. *Zh. Obsch. Khim.* (1991) **61** (3), 657–66.

[200] Zabirov N.G., Soloviev V.N., Shamsevaleev F.M., Cherkasov R.A. and Martinov I.V. *Zh. Obsch. Khim.* (1991) **61** (8), 1748–54.

[201] Zabirov N.G., Shamsevaleev F.M. and Cherkasov R.A. *Zh. Obsch. Khim.* (1992) **62** (5), 1071–8.

[202] Vetrova Z.P., Zabirov N.G., Shuvalova T.N., Smerkovich L.S., Ivanova L.A., Golberg A.K. and Karabanov N.T. *Zh. Obsch. Khim.* (1992) **62** (5), 1079–83.

[203] Gratchev M. *Phosphorus Sulfur* (1996) **109–110**, 417–20.

[204] Roman J.S., Levenfeld B., Madruga E.L. and Varion J.P. *J. Polymer. Sci., A — Polym. Chem.* (1991) **29** (7), 1023–30.

[205] Roman L.C. and Levenfeld B. *Makromol.* (1991) **24** (23), 6083–8.

[206] Chung K.N., Kim K.M., Kim J.N. and Ryn E.K. *Synth. Commun.* (1991) **21** (18-1), 1917–22.

[207] Zins A.B., Wyse D.L. and Koskinen W.C. *Weed Sci.* (1991) **39** (2), 262–9.

[208] Frank R., Clegg B.S. and Patni N.K. *Arch. Environ. Contamin. Toxicol.* (1991) **21** (2), 253–62.

[209] Lewandowski M., Chui Y.C., Levi P. and Hodgson E. *Toxicol. Lett.* (1991) **55** (2), 223–31.

[210] Antonov G., Zaicov C., Bouzidi A., Mitova S., Michaelova A., Halkova J. and Choumkov N. *J. Toxicol. Clin. Experim.* (1991) **11** (6), 349–56.

[211] Sahide I.B. and Wei C.C. *Bull. Environ. Contamin. Toxicol.* (1993) **50** (1), 24–28.

[212] Negre M., Gennari M., Raimondo E., Celi L., Trevisan M. and Capri E. *J. Agricult. Food Chem.* (1992) **40** (6), 1071–5.

[213] Sorci J.J. and Macalady D.L. *J. Agricult. Food Chem.* (1993) **41** (10), 1760–6.

[214] Ademola J.I., Wester R.C. and Maibach H.I. *Toxicol. Appl. Pharmacol.* (1993) **121** (1), 78–86.

[215] Efremov D.A., Tebby J.C. and Zavlin P.M. *Phosphorus Sulfur* (1996) **109–110**. 465–8.

[215b] Oberlander E.A. and Tebby J.C. *Heteroatom Chemistry* (1998) **9** (2) 261–3.

[216] Zavlin P.M., Efremov D.A. and Tebby J.C. Symposium on Organic Chemistry. St Petersburg, 1995, Abstracts, p. 127.

[217] Efremov D.A. and Zavlin P.M. *Zh. Obsch. Khim.* (1997) **67** (2), 228–35.

[218] Zavlin P.M. and Efremov D.A. *Zh. Obsch. Khim.* (1997) **67** (6), 932–7.

[219] Gasteiger J., Hondelmann U., Rose P. and Witzenbichler W. *J. Chem. Soc. Perkin Trans. 2* (1995) (2), 193–204.

[220] Henrichs P.M., Estep K., Musza L.L. and Rodgers C.A. *J. Am. Chem. Soc.* (1995) **117** (7), 2058–66.

[221] Zaugg H.E. and Martin W.B. *Org. Reactions.* (1965) **14**, 52–269.

[222] Valentijn A.R.P.M., Van der Marel G.A., Cohen L.N. and Van Boom J.H. *Synlett.* (1991) (9), 663–4.

[223] Mizrahi V. and Modro T.A. *J. Org. Chem.* (1983) **48** (18), 3030–7.

[224] Efremov D.A. and Zavlin P.M. *XI Internat. Conf. on Chemistry of Phosphorus Compounds.* Kazan, 1996, Abstracts, p. 161.

[225] Efremov D.A. and Zavlin P.M. *Zh. Obsch. Khim.* (1997) **67** (6), 932–7.

[226] Gautier J.-A., Miocque M. and Farnoux C. C. "Preparation and Synthetic Uses of Amidines" in *The Chemistry of Amidines and Imidates,* ed. S. Patai, Wiley, London, 1975, pp. 283–348.

[227] Tebby J.C., Zavlin P.M. and Efremov D.A. *Zh. Obsch. Khim.* (1994) **64** (2), 242–5.

[228] Efremov D.A., Oberlander E.A. and Tebby J.C. *Phosphorus Sulfur.* (1994) **86**, 81–3.

[229] Zavlin P.M., Tebby J.C., Efremov D.A. and Oberlander E.A. *Zh. Obsch. Khim.* (1993) **63** (9), 2017–9.

[230] Efremov D.A., Oberlander E.A., Tebby J.C., Zavlin P.M. and Gribanov A.V. *J. Chem. Soc. Perkin Trans. 1*, (1994) 2443–7.

[231] Wislicenus W. and Goldschmidt M. *Ber.* (1900) **33**, 1467–71.

[232] Mumm O., Hesse H. and Volquartz H. *Ber.* (1915) **48**, 379–91.

[233] Curtin D.Y. and Miller L.L. *J. Am. Chem. Soc.* (1967) **89** (3), 637–47.

[234] Curtin D.Y. and Miller L.L. *Tetrahedron Lett.* (1965) 1869–76.

[235] Schulenberg J.W. and Archer S. *Organic Reactions,* ed. S. Patai, Wiley, London, 1965, **14**, pp. 1–51.

[236] Schwarz J.S.P. *J. Org. Chem.* (1972) **37** (18), 2906–8.

[237] McCormack M.T. and Hegarty A.F. *J. Chem. Soc. Perkin Trans. 2* (1976) 1701–9.

[238] McCarthy D.G. and Hegarty A.F. *J. Chem. Soc. Perkin Trans. 2* (1977) 1080–4.

[239] McCarthy D.G. and Hegarty A.F. *J. Chem. Soc. Perkin Trans. 2* (1977) 1085–94.

[240] Hegarty A.F. and Mc Cormack M.T. *J. Chem. Soc. Perkin Trans 2* (1976) 1701–9.

[241] Curtin D.Y. and Engelmann J.H. *Tetrahedron Lett.* (1968) 3911–3.

[242] Curtin D.Y. and Engelman J.H. *J. Org. Chem.* (1972) **37** (22), 3439–43.

[243] Challis B.C., Challis J.A. and Mc Dermott I.R. *J. Chem. Soc. Perkin Trans. 2* (1979) 634–41.

[244] Chapman A.W. *J. Chem. Soc.* (1925) 1992–8.

[245] Chapman A.W. *J. Chem. Soc.* (1926) 2296–300.

[246] Chapman A.W. *J. Chem. Soc.* (1927) 1743–51.

[247] McCarty C.G. "syn-anti Isomerization and rearrangements" in *The Chemistry of the Carbon–Nitrogen Double Bond*, Wiley, London, 1970, pp. 363–464.

[248] McCarty C.G. and Garner L.A. "Rearrangements involving imidic acid derivatives" in *The Chemistry of Amidines and Imidates*, ed. S. Patai, Wiley, London, 1975, pp. 189–240.

[249] Wiberg K.B. and Rowland B.I. *J. Am. Chem. Soc.* (1955) **77** (8), 2205–9.

[250] Arbuzov A.E and Shishkin V.E. *Dokl. Akad. Nauk USSR* (1961) **141** (2), 349–52.

[251] Arbuzov A.E. and Shishkin V.E. *Zh. Obsch. Khim.* (1964) **34** (11), 3579–82.

[252] Challis B.C. and Frenkel A.D. *J. Chem. Soc. Perkin Trans. 2* (1978) 192–7.

[253] Mumm O. and Moller F. *Ber.* (1937) **70** (11), 2214–27.

[254] Beak P., Lee J.-K. and Mc Kinnie B.G. *J. Org. Chem.* (1978) **43** (7), 1367–72.

[255] Cramer F. and Hennrich N. *Chem. Ber.* (1961) **94**, 976–89.

[256] Moodie R.B., Thomas P.N. and Schofield K. *J. Chem. Soc. Perkin Trans. 2* (1977) 1693–705.

[257] Carvalho E., Norberto F., Rosa E., Iley J. and Patel P. *J. Chem. Res. (S)* (1985) 132–3.

[258] Carvalho E., Iley J. and Rosa E. *J. Chem. Soc., Chem. Commun.* (1988) 1249–50.

[259] Carvalho E., Iley J., Norberto F. and Rosa E. *J. Chem. Res. (S)* (1989) 260–1.

[260] Iley J., Carvalho E., Norberto F. and Rosa E. *J. Chem. Soc. Perkin Trans. 2* (1992) 281–9.

[261] Norman R.O.C. and Radda G.K. *J. Chem. Soc.* (1961) 3030–7.

[262] Knowles J.R. and Norman R.O.C. *J. Chem. Soc.* (1961) 3888–91.

[263] Hartshorn S.R., Moodie R.B. and Schofield K. *J. Chem. Soc. (B)* (1971) 2454–61.

[264] Hartshorn S.R., Hoggett J.G., Moodie R.B., Schofield K. and Thompson M.J. *J. Chem. Soc. (B)* (1971) 2461–9.

[265] Lynch B.M., Chen C.M. and Wigfield Y.-Y. *Canad. J. Chem.* (1968) **46**, 1141–8.

[266] Hardy F.E. and Robson P. *J. Chem. Soc. (B)* (1967) 1151–4.

[267] Brook A.G. and Bassindale A.R. "Molecular Rearrangements of Organosilicon Compounds" in *Organic Chemistry*, Vol. 42, *Rearrangements in Ground and Excited States*, Vol. 2, ed. P. de Mayo, Academic Press, New York, 1980, pp. 193–227.

[268] Kirsanov A.V. and Makitra R.G. *Zh. Obsch. Khim.* (1958) **28** (1), 35–9.

[269] Kolodiazhni O.I., Iakovlev V.N. and Kukhar V.P. *Zh. Obsch. Khim.* (1979) **49** (11), 2170–7.

[270] Pudovik A.N., Cherkasov R.A., Zimin M.G. and Zabirov N.G. *Izv. Akad. Nauk. USSR. Ser. Khim.* (1979) 4, 861–6.

[271] Zimin M.G., Kamalov R.M., Cherkasov R.A. and Pudovik A.N. *Phosphorus Sulfur* (1982) **13**, 371–8.

[272] Onisko P.P., Kim T.V., Kiseleva E.I., Prokopenko V.P. and Sinitsa A.D. *Zh. Obsch. Khim.* (1988) **58** (1), 35–41.

[273] Onisko P.P., Suvalova E.A., Chudakova T.I., Kim T.V., Kiseleva E.I. and Sinitsa A.D. *Zh. Obsch. Khim.* (1989) **59** (9), 2138–9.

[274] Onisko P.P., Suvalova E.A., Chudakova T.I. and Sinitsa A.D. *Zh. Obsch. Khim.* (1991) **61** (4), 843–7.

[275] Zimin M.G., Burilov A.R. and Pudovik A.N. *Zh. Obsch. Khim.* (1981) **51** (2), 469–70.

[276] Pudovik A.N. and Zimin M.G. *Uspekhi Khim.* (1983) **52** (11), 1803–30.

[277] Derkach G.I., Shokol V.A. and Kirsanov A.V. *Zh. Obsch. Khim.* (1960) **30** (10), 3393–7.

[278] Kabachnik M.I., Gilarov V.A., Chjan Chjen-de and Matrosov E.I. *Izv. Akad. Nauk. USSR. Ser. Khim.* (1962) **9**, 1589–99.

[279] Zimin M.G., Zabirov N.G., Kamalov R.M. and Pudovik A.N. *Zh. Obsch. Khim.* (1986) **56** (12), 2660–6.

[280] Zimin M.G., Kamalov R.M., Afanasev M.M. and Pudovik A.N. *Zh. Obsch. Khim.* (1986) **56** (12), 2666–73.

[281] Kirsanov A.V., Derkach G.I. and Makitra R.G. *Zh. Obsch. Khim.* (1958) **28** (5), 1227–32.

[282] Derkach G.I., Lepesa A.M. and Kirsanov A.V. *Zh. Obsch. Khim.* (1961) **31** (10), 3424–33.

[283] Shokol V.A. and Derkach G.I. *Zh. Obsch. Khim.* (1965) **35** (8), 1468–71.

[284] Baraniak J. and Stec W.J. *Tetrahedron Lett.* (1991) **32**, 137–41.

[285] Baraniak J. and Stec W.J. *Tetrahedron Lett.* (1991) **32**, 4193–6.

[286] McEnvoy F.J., Greenblatt E.N., Osterberg A.C. and Allen Jr. G.R. *J. Med. Chem.* (1970) **13**, 295–7.

[287] Mukherjee R. and Block P. *J. Chem. Soc. C.* (1971) 1596–600.

[288] Westby T.R. and Barfknecht C.F. *J. Med. Chem.* (1973) **16** (1), 40–3.

[289] Goldberg V.D. and Harris M.M. *J. Chem. Soc. Perkin Trans. 2.* (1973), 1303–6.

[290] Pratt R.F. and Bruice T.C. *Biochemistry* (1971) **10** (17), 3178–85.

[291] Clark V.M., Hutchinson D.W., Kirby A.J. and Warren S.G. *Angew. Chem.* (1964) **16**, 704–12.

[292] Krasnov K.A., Slesarev V.I. and Artemeva Z.L. *Zh. Obsch. Khim.* (1989) **25** (7), 1553–7.

[293] Salikhov I.Sh., Reznik V.S., Shvetsov Yu.S. and Shagidullin R.R. Jr. *Izv. Akad. Nauk USSR* (1986) **10**, 2283–9.

[294] Pestov D.V., Slesarev V.I., Ginak A.I. and Slesareva V.I. *Khim. Geterocycl. Soedin.* (1988) **7**, 952–6.

[295] Goodman L.G. and Gilman A.G. *Pharmacological Basis of Therapeutics*, Macmillan Publishers Co, NY, 1975, pp. 102–23.

[296] Jovanovich M.V. and Biehl E.R. *Heterocycles* (1986) **24** (11), 3129–41.

[297] Troschutz R. and Anders E. *Arch. Pharm.* (1992) **325** (6), 341–8.

[298] Zavlin P.M., Efremov D.A. and Govorkov A.L. *Zh. Obsch. Khim.* (1992) **62** (1), 220–1.

[299] Balakina M.Yu. and Zuev M.B. *Phosphorus Sulfur* (1996) **111**, 134.

[300] Zavlin P.M., Efremov D.A. and Govorkov A.L. *Zh. Obsch. Khim.* (1992) **62** (1), 221–2.

[301] Zavlin P.M., Diakonov A.N., Kuznetsov L.L., Efremov D.A. and Belous V.M. *Zh. Nauchnoi Prikl. Photogr.* (1996) **41** (2), 66–70.

[302] Coe D.G., Perry B.J. and Brown R.K. *J. Chem. Soc.* (1957) 3604–7.

[303] Edmundson R.S. *J. Chem. Soc., C* (1967) 1635–7.

[304] Jones R.A.Y., Katritzky A.R. and Michalski J. *Proc. Chem. Soc.* (1959) 321.

[305] Michalski J., Mikolajczyk M. and Skowronska A. *Chem. Ind.* (1962) **18**, 1053–4.

[306] Michalski J., Mikolajczyk M. and Ratajczak A. *Chem. Ind.* (1962) **18**, 819–20.

[307] Michalski J. *Bull. Soc. Chim. France* (1963) **1**, 11–16.

[308] Michalski J., Mikolajczyk M. and Ratajczaka A. *Bull. Acad. Polon. Sci., Ser. Sci. Chem.* (1965) **13** (4), 277–84.

[309] Michalski J., Mikolajczyk M. and Mlotkowska B. *Chem. Ber.* (1969) **102** (1), 90–4.

[310] Michalski J. and Skowronska A. *J. Chem. Soc., C* (1970) 703–7.

[311] Zavlin P.M. and Efremov D.A. *Zh. Obsch. Khim.* (1996) **66** (1), 165.

[312] Onisko P.P., Kim T.V., Kiseleva E.I. and Sinitsa A.D. *Zh. Obsch. Khim.* (1987) **57** (6), 1233–40.

[313] Fetukhin V.N. and Vovk M.V. *Zh. Obsch. Khim.* (1983) **53**, (8), 1763–7.

[314] Onisko P.P., Suvalova E.A., Chudakova T.I. and Sinitsa A.D. *Zh. Obsch. Khim.* (1990) **60** (7), 1554–8.

[315] Onisko P.P., Kiseleva E.I., Kim T.V. and Sinitsa A.D. *Zh. Obsch. Khim.* (1985) **55** (2), 454–6.

[316] Rondeau R.E., Rosenberg H.M. and Dunbar D.J. *J. Mol. Spectr.* (1969) **29** (3), 305–11.

[317] Roumestant M.L., Viallefont P., Elguero J. and Lacquier R. *Tetrahedron Lett.* (1969) 495–8.

[318] Chen B.C., von Philipsborn W. and Nagarajan K. *Helv. Chim. Acta.* (1983) **66** (146), Fasc. 5, 1537–55.

[319] Nesmeyanov A.N., Zavelovich E.B., Babin V.N., Kochetkova N.S. and Fedin E.I. *Tetrahedron* (1975) **31**, 1461–2.

[320] Claramunt R.M., Elguero J., Marzin C. and Seita J. *Anal. Quim.* (1979) **75**, 701–6.

[321] Katritzky A.R., Rachwal S. and Hitchings G.J. *Tetrahedron* (1991) **47**, 2683–732.

[322] Catalan J., Sanchez-Cabezudo M., de Paz J.L.G., Elguero J., Taft R.W. and Anvia F. *J. Comput. Chem.* (1989) **10** (3), 426–33.

[323] Bernardini A., Viallefont P., Gelize-Duvigneau M. and Sauvaitre H. *J. Mol. Struct.* (1976) **34** (2), 245–62.

[324] Fayet J.-P., Vertut M.-C., Mauret P., Faure R., Galy J.-P., Vincent E.-J. and Elguero J. *C. R. Acad. Sc. Paris. Serie C.* (1976) **283** (5), 157–8.

[325] Starova G.L., Frank-Kamenetskaya O.V., Makarski V.V. and Lopirev B.A. *Crystallogr.* (1978) **23** (4), 849–51.

[326] Sauvaitre H., Teysseyre J. and Elguero J. *Bull. Soc. Chim. France* (1976) 3–4. 635–41.

[327] Escande A., Lapasset J., Faure R., Vincent E.-J. and Elguero J. *Tetrahedron.* (1974) **30**, 2903–9.

[328] Schroeder M.A., Makino R.C. and Tolles W.M. *Tetrahedron.* (1973) **29**, 3463–8.

[329] Tornkvist C., Bergman J. and Liedberg B. *J. Phys. Chem.* (1991) **95** (8), 3123–8.

[330] Katritzky A.R., Yannakopoulou K., Anders E., Stevens J. and Szafran M. *J. Org. Chem.* (1990) **55** (22), 5683–7.

[331] Fabian W.M.F. *Z. Naturforsch.* (1990) **45à**, 1328–34.

[332] Anders E., Katritzky A.R., Malhotra N. and Stevens J. *J. Org. Chem.* (1992) **57** (13), 3698–705.

[333] Ritchie J.P. *J. Org. Chem.* (1989) **54** (15), 3553–60.

[334] Claramunt R.M., Elguero J. and Nye A.F.M.J. *Afinidad.* (1977) **34** (350), 545–51.

[335] Abu-Eittan R.H., Hamed M.M., Nigm A. and El-Azhary A. *Int. J. Quant. Chem.* (1985) **28** (6), 731–40.

[336] Theocharis A.B., Alexandrou N.E., Papadopoulos M.G. and Waite J. *J. Chem. Res. (S)* (1991) **4**, 104–5.

[337] Volkov V.V., Zavlin P.M., Dronov M.M., Matevosian G.L. and Maksimov I.B. *Symp. on Ophthalmol. Moscow*, 1982, Abstracts, p. 24.

[338] Matevosian G.L. and Zavlin P.M. *VII Vsesouzn. Conf. Organophosph. Comp.*, Leningrad, Nauka, 1987, pp. 216–22.

[339] Brukshtus A.B., Garalene V.N., Sirvidite A.P.-P. and Daukshas V.K. *Chim.-Pharm. Zh.* (1994) **28** (6), 24–6.

[340] Denny W.A., Rewcastle G.W. and Bangley B.C. *J. Med. Chem.* (1990) **33**, 814–9.

[341] Bru-Magniez N., Gungor T., Lacrampe J., Launay M. and Teulon J.M. *Europ. Pat.* 385,850 (1990).

[342] Mertens A., Muller-Beckmann B., Kampe W., Holk J.-P. and von der Saal W. *J. Med. Chem.* (1986) **30**, 1279–88.

[343] Declerq J.P., Dubourg A., Etemad-Moghadam G., Klaebe A. and Perie J.J. *Phosphorus Sulfur* (1984) **21**, 31–7.

[344] Brosalina E.B. and Grachov S.A. *Bioorg. Khim.* (1986) **12**, 248–56.

[345] Matevosian G.L. and Zavlin P.M. *Zh. Obsch. Khim.* (1982) **52** (7), 1441–54. Matevosian G.L. and Zavlin P.M. Pat. 732,272. USSR (1979). Sovetkina V.E., Shishov A.D., Matevosian G.L. and Zavlin P.M. Pat. 959,724. USSR (1982).

[346] Razvodovskaya L.V., Grapov A.F. and Melnikov N.N. *Uspekhi Khim.* (1982) **51** (1), 239–65.

[347] *Scientific Basis of Plants Protection*, ed. Fadeev Yu.N. and Novozhilova K.V. Kolos, Moscow, 1984, p. 138.

[348] Matevosian G.L. and Zavlin P.M. *Zh. Obsch. Khim.* (1979) **49** (6), 1252–5.

[349] Matevosian G.L. and Zavlin P.M. *Khim. Geterocycl. Soed.* (1990) 6, 723–40.

[349a] Matevosian G.L. and Zavlin P.M. *14th Int. Conf. on Phosphorus Chemistry, Cincinnati, 1998* Abstracts, pp. LH5–6.

[350] Schroeder M.A., Makino R.C. and Tolles W.M. US Army Aberdeen Research and Development Center, Maryland, 1972, p. 9.

[351] Katritzky A.R. and Lan X. *Chem. Soc. Rev.* (1994) **23** (6), 363–73.

[352] Breslav Yu.A., Terentev E.G., Kagakin E.I., Ushanov G.G. and Vlasov V.G. *The Stable Forms of Silver Halides Microcrystals*, VNIITEKHIM, Moscow, 1988.

[353] Haque M.R. and Rasmussen M. *Austr. J. Chem.* (1994) **47** (8), 1523–35.

[354] Zavlin P.M. and Efremov D.A. *Zh. Nauchn. i Prikl. Photograph.* (1992) **2**, 128–32.

[355] Zavlin P.M., Efremov D.A. and Tebby J.C. *Zh. Obsch. Khim.* (1994) **64** (4), 692–3.

[356] Efremov D.A., Tebby J.C. and Zavlin P.M. *Phosphorus Sulfur* (1994) **92**, 167–70.

[357] Zavlin P.M., Efremov D.A. and Gugkaev A.V. *Zh. Obsch. Khim.* (1990) **60** (5), 1182–3.

[358] Soborovski L.Z., Zinoviev Yu.M. and Spiridonova T.G. *Zh. Obsch. Khim.* (1959) **29** (11), 3592–7.

[359] Petrov K.A. II Conf. Khim. Prim. Phosphororg. Soed. Moscow. Akad. Nauk USSR, 1962, p. 125.

[360] Petrov K.A., Bliznuk N.K. and Savostenok V.A. *Zh. Obsch. Khim.* (1961) **31** (4), 1361–6.

[361] Cook H.G., Ilett J.D., Saunders B.C., Stacey G.J., Watson H.G., Wilding I.G.E. and Woodcock S.J. *J. Chem. Soc.* (1949) 2921–7.

[362] McCombie H. and Saunders B.C. *Nature* (1946) **157**, 776–7.

[363] Ivanovski V.I. *Chemistry of Heterocyclic Compounds*, Visshaya Shkola, Moscow, 1976, p. 559.

[364] Guibe-Jampel E., Wakselman M. and Vilkas M. *Bull. Soc. Chem. France* (1971) 4, 1308–14.

[365] Atherton F.R., Openshaw N.T. and Todd A.R. *J. Chem. Soc.* (1945) 660–3.

[366] Anand N. and Todd A.R. *J. Chem. Soc.* (1951) 1867–72.

[367] Atherton F.R. and Todd A.R. *J. Chem. Soc.* (1947) 674–8.

[368] Zavlin P.M. and Matevosian G.L. *Zh. Obsch. Khim.* (1977) **47** (2), 483.

[369] *General Organic Chemistry*, ed. D. Barton and D. Ollis Russian ed, Khimia, Moscow, 1983, **5**, p. 52.

[370] Nikolaenko L.N. and Degtiarev E.V. *Zh. Obsch. Khim.* (1967) **37** (6), 1350–2.

[371] Ramierez F. and Marecek J.F. *Tetrahedron Lett.* (1977) **9** (11), 967–9.

[372] Ramierez F., Okazaki H. and Marecek J.F. *Synthesis* (1975) 9, 637–8.

[373] Ramierez F., Ricci J.S., Okazaki H., Marecek J.F. and Lewy M. *Phosphorus Sulfur* (1984) **20** (3), 279–99.

[374] Mukaiyama T. and Hashimoto M. *Bull. Chem. Soc. Japan* (1971) **44** (8), 2284.

[375] Hultquist D.E. *Biochim. Biophys. Acta* (1988) **153**, 329–38.

[376] Zavlin P.M., Efremov D.A. and Tebby J.C. *Zh. Obsch. Khim.* (1994) **64** (8), 1306–7.

[377] Yamamoto K. and Watanabe H. *Chem. Lett.* (1982) 1225–8.

[378] Perry R.J. and Wilson B.D. *J. Org. Chem.* (1993) **58** (25), 7016–21.

[379] Sykes B.M., Antwell G.J., Denny W.A., McLennan D.J. and O'Connor C.J. *J. Chem. Soc. Perkin Trans.* 2 (1995) 337–42.

[380] Hein D.W., Alheim R.J. and Leavitt J.J. *J. Am. Chem. Soc.* (1957) **79**, 427–29.

[381] Grunze H. *Z. Anorg. Chem.* (1958) **296**, 63–72.

[382] Groenweghe L.C.D., Payne J.H. and Van Wazer J.R. *J. Am. Chem. Soc.* (1960) **82** (20), 5305–11.

[383] Grunze H. *Z. Anorg. Allg. Chem.* (1958) **296**, 63–72.

[384] Grunze H. *Angew. Chem.* (1959) **71**, 70.

[385] Grunze H. *Angew. Chem.* (1959) **71**, 407.

[386] Grunze H. *Chem. Ber.* (1959) 92.

[387] Grunze H. *Chem. Ber.* (1959) 850.

[388] Woodstock W.H. US Pat. 2,402,703 (1946).

[389] Balarev D. *Z. Anorg. Chem.* (1914) **88**, 145–9.

[390] Berger H. *Monatsh.* (1970) **101**, 559–67.

[390a] Van Wazer J.R. *J. Am. Chem. Soc.* (1950) **72** (2), 644–7.

[391] Ketelaar J.A.A. and Bloksma A.H. *Rec. Trav. Chim.* (1948) **67** (11), 665–76.

[392] Hall S.A. and Jacobson M. *Ind. Eng. Chem.* (1948) **40** (4), 694–9.

[393] Mizerakh L.N., Evdakov V.P. and Sandalova L.Yu. *Zh. Obsch. Khim.* (1965) **35** (10), 1871–6.

[394] Schwarzmann E. and Van Wazer J.R. *J. Am. Chem. Soc.* (1960) **82** (23), 6009–13.

[395] Montignie W.C. *Bull. Soc. Chim. France* (1947) 336–7.

[396] Smith W.C. *J. Am. Chem. Soc.* (1960) **82**, 3835–8.

[397] Lecher H.Z., Chao T.H., Whitehouse K.C. and Greenwood R.A. *J. Am. Chem. Soc.* (1954) **74** (4), 1045–51.

[398] Lecher H.Z., Chao T.H. and Whitehouse K.C. US Pat. 2,717,906 (1955).

[399] Lecher H.Z., Greenwood R.A, Whitehouse K.C. and Chao T.H. *J. Am. Chem. Soc.* (1956) **78** (19), 5018–22.

[400] Greenwood R.A., Scalera M. and Lecher H.Z. US Pat. 2814645 (1957).

[401] Lecher H.Z. and Chao T.H. Pat. 963,425. Ger. (1960).

[402] Petrov K.A., Chizhov V.M., Pokatun V.P. and Agafonov S.V. *Uspekhi Khim.* (1986) **55** (11), 1834–53.

[402a] Lennon P., Vulfson S., d'Avignon A., Rath N. and Gard J. *14th Int. Conf. on Phosphorus Chemistry, Cincinnati, 1998*, Abstracts, pp. LH5–3.
Lennon P.J. and Vulfson S.G. Pat. WO 9829106 (1998).

[403] Sauer R.O. *J. Am. Chem. Soc.* (1944) **66** (10), 1707–10.

[404] Turner A., Honn F.J. and Mark H. *J. Polymer. Sci.* (1946) **1** (2), 102–20.

[405] Orlov N.F. and Voronkov M.G. *Zh. Obsch. Khim.* (1960) **30** (7), 2223–9.

[406] Voronkov M.G. *Zh. Obsch. Khim.* (1955) **25** (3), 469–70.

[407] Feher F., Kuhlborsch G., Blumcke A., Keller H. and Lippert K. *Chem. Ber.* (1957) **90** (1), 134–44.

[408] Mileshkevich V.P. and Karlin A.V. *Zh. Obsch. Khim.* (1970) **40** (12), 2573–7.

[409] Borisov S.N., Voronkov M.G. and Lukevits E.Ia. *Organosilicon Derivatives of Phosphorus. Part 1*, Khimia, Leningrad, 1968.

[410] Imamoto T., Matsumoto T., Yokoyama H., Yokoyama M. and Yamaguchi K. *J. Org. Chem.* (1984) **49** (6), 1105–10.

[411] Kreshkov A.P. and Karateev D.A. *Chemistry and Application of Organosilicon Compounds*, II Conf. Akad. Nauk USSR, Moscow, 1962, pp. 324–8.

[412] Mizhiritski M.D. and Reikhsfeld V.O. *Zh. Obsch. Khim.* (1985) **55** (8), 1883–4.

[413] Orlov N.F. and Voronkov M.G. *Zh. Obsch. Khim.* (1962) **32** (4), 608.

[414] Gavrilin G.F. and Vovsi B.A. Pat. 168,443. USSR (1965).

[415] Chivikova A.N., Kreshkov A.P., Darashkevich M.L., Mishliaeva L.V., Karateev D.A. and Khananashvili L.M. *Chemistry and Application of Organosilicon Compounds*, 1st ed, Leningrad, 1958, pp. 178–84.

[416] Kreshkov A.P. and Karateev D.A. *Zh. Prikl. Khim.* (1959) **32** (2), 369–74.

[417] Kreshkov A.P. and Karateev D.A. *Zh. Prikl. Khim.* (1960) **33** (2), 413–7.

[418] Khananashvili L.M., Chivikova A.N., Kreshkov A.P. and Darashkevich M.L. *Chemistry and Application of Organosilicon Compounds*, 6th ed, Akad. Nauk USSR, Leningrad, 1961, pp. 156–61.

[419] Kreshkov A.P. and Karateev D.A. Pat. 127,262. USSR (1959).

[420] Khramova V.I. and Kreshkov A.P. *Chemistry and Application of Organosilicon Compounds*, 4th ed., Leningrad, 1958, pp. 101–6.

[421] Kreshkov A.P. and Karateev D.A. *Zh. Obsch. Khim.* (1959) **29** (12), 4082–5.

[422] Kreshkov A.P. and Karateev D.A. and Furst V. *Zh. Prikl. Khim.* (1961) **34** (12), 2711–6.

[423] Kreshkov A.P. and Karateev D.A. *Zh. Prikl. Khim.* (1957) **30** (9), 1416–21.

[424] Kreshkov A.P., Khramova V.I. and Karateev D.A. *The Methods of Improving Cements Solutions*, 27th ed, Moscow Technological Institute, Moscow, 1959, pp. 306–14.

[424a] O'Lenick A.J. US Pat. 5,070,171 (1991).

[425] Nizamov I.S., Kuznetsov V.A., Batieva E.S. and Pudovik A.N. Pat. 1,735,300. Russia (1992).

[426] Nizamov I.S., Kuztnetsov V.A., Batieva E.S., Alfonsov V.A. and Pudovik A.N. *Izv. Acad. Sc. Russia. Ser. Khim.* (1992) 8, 1933–9.

[427] Nizamov I.S., Kuztnetsov V.A., Batieva E.S., Alfonsov V.A. and Pudovik A.N. *Zh. Obsch. Khim.* (1992) **62** (7), 1665–6.

[428] Kempe K.D. and Fisher E. US Pat. 3,642,956 (1972).

[429] Rinse J. US Pat. 2,886,586 (1959).

[430] Schmidt M. and Ruidisch I. *Chem. Ber.* (1962) **95**, (6), 1434–7.

[431] Nizamov I.S., Kuztnetsov V.A., Batieva E.S. and Alfonsov V.A. *Izv. Acad. Sc. Russia. Ser. Khim.* (1992) 10, 2457–8.

[432] Nizamov I.S., Kuztnetsov V.A., Batieva E.S., Alfonsov V.A. and Pudovik A.N. *Phosphorus Sulfur* (1993) **79** (1–4), 179–85.

[433] Nizamov I.S., Kuztnetsov V.A., Batieva E.S., Alfonsov V.A. and Pudovik A.N. *Heteroatom Chem.* (1993) **4** (4), 379–82.

[434] Nizamov I.S., Kuztnetsov V.A., Batieva E.S., Alfonsov V.A. and Pudovik A.N. *Zh. Obsch. Khim.* (1994) **64** (8), 1333–6.

[435] Nizamov I.S., Kuztnetsov V.A. and Batieva E.S. *Phosphorus Sulfur* (1994) **90**, 139–47.

[436] Nizamov I.S., Kuztnetsov V.A., Batieva E.S. and Pudovik A.N. *Zh. Obsch. Khim.* (1996) **66** (3), 518–9.

[437] Plesch P.H. *High Vacuum Techniques for Chemical Synthesis and Measurements*, Cambridge University Press, Cambridge, 1989, pp. 129, 140.

[438] Kipping F.S. *J. Chem. Soc.* (1890) **47–48**, 980–8.

[439] Diels O. and Meyerheim G. *Ber.* (1907) **40**, 355–63.

[440] Lacey R.N. Ketenes in *The Chemistry of Alkenes*. Interscience Publishers, London, New York, 1964, pp. 1161–227.

[441] Pauw E.A. *Rec. Trav. Chim.* (1936) **55** (3), 215–26.

[442] Volkenshtein M.V. *Uspekhi Khim.* (1935) **4** (4), 610–31.

[443] Roberts C.R. and De Puy C.H. *Zh. Obsch. Khim.* (1989) **59** (10), 2153–61.

[444] Allen A.D., McAllister M.A. and Tidwell T.T. *J. Chem. Soc. Chem. Commun.* (1995) 2547–8.

[445] Huntress Jr. W.T., Allen M. and Delitsky M. *Nature* (1991) **352**, 316.

[446] Staudinger H. and Bereza S. *Ber.* (1908) **41**, 4461.

[447] Staudinger H. *Ber.* (1911) **44**, 533–43.

[448] Krugerke T., Buschmann J., Kleemann G., Luger P. and Seppelt K. *Angew. Chem. Int. Ed. Engl.* (1987) **26** (8), 799–801.

[449] Bittner J. and Seppelt K. *Chem. Ber.* (1990) **123**, 2187–90.

[450] Haas A. and Praas H.-W. *Chem. Ber.* (1992) **125**, 571–9.

[451] Krugerke T. and Seppelt K. *Chem. Ber.* (1988) **121**, 1977–81.

[452] Dorsey D.A., King S.M. and Moore H.W. *J. Org. Chem.* (1986) **51** (14), 2814–6.

[453] Kim Y.K., Grindahl G.A., Greenwald J.R. and Pierce O.R. *J. Heterocycl. Chem.* (1974) **11** (4), 563–8.

[454] Stevens C.L. and Singhal G.H. *J. Org. Chem.* (1964) **29** (1), 34–37.

[454a] Stevens C.L., Singhal G.H. and Ash A.B. *J. Org. Chem.* (1967) **32** (9), 2895.

[455] Krow G.R. *Angew. Chem. Internat. Ed.* (1971) **10** (7), 435–49.

[456] *Comprehensive Organic Chemistry. The Synthesis and Reactions of Organic Compounds. Vol. 2. Nitrogen Compounds, Carboxylic Acids, Phosphorus Compounds*, Pergamon Press, Oxford, 1979, p. 516.

[457] Sander M. *Monatch. Chem.* (1964) **95** (2), 608–16.

[458] Shvedova V.N., Dolgopolski I.M. and Diachishina L.M. *Zh. Obsch. Khim.* (1969) **39** (4), 772–5.

[459] Sullivan R. *J. Org. Chem.* (1969) **34** (6), 1841–6.

[460] Volkov A.N. and Nikolskaya A.N. *Uspekhi Khim.* (1977) **46** (4), 712–39.

[461] Alekseev Yu.I. and Pashkevich D.S. *Zh. Prikl. Khim.* (1993) **66** (9), 2031–7.

[462] Alekseev Yu.I. and Pashkevich D.S. *Zh. Prikl. Khim.* (1993) **66** (10), 2354–8.

[463] Alekseev Yu. I., Koroliov V.L. and Pashkevich D.S. *Zh. Prikl. Khim.* (1994) **67** (11), 1830–4.

[464] Nakajima N. and Ubukata M. *Tetrahedron Lett.* (1997) **38**, 2099–102.

[465] Kent R.E. and McElvain S.M. *Org. Synth. Collect. Vol. 3*, Wiley, New York, 1955, pp. 493–4.

[466] Fieser M. and Fieser L. *Reagents for Organic Synthesis. Vol. 2*, Wiley, New York, 1969, pp. 329–30.

[467] Teague P.C. and Short W.A. *Org. Synth. Collect. Vol. 4*, Wiley, New York, 1963, pp. 706–7.

[468] McElvain S.M. and Clarke R.L. *J. Am. Chem. Soc.* (1947) **64**, 2661–3.

[469] Snyder H.R. and Elston C.T. *J. Am. Chem. Soc.* (1954) **76**, 3039–40.

[470] Hauser C.R. and Murray J.G. *J. Am. Chem. Soc.* (1955) **77**, 2851–2.

[471] Merck E. Pat. 670,131 *Ger. Chem. Abstr.* (1939) **33**, 2909.

[472] Schwarz G. *Org. Synth. Collect. Vol. 3*, Wiley, New York, 1955, pp. 332–3.

[473] Bischler A. and Napieralski B. *Ber.* (1893) **26**, 1903.

[474] Waley W.M. and Govindachari T.R. *Org. Reactions* (1951) **6**, 74–99.

[475] Morrison G.C., Cetenko W. and Shavel J., Jr. *J. Org. Chem.* (1964) **29**, 2771–2.

[476] Hassner A. and Stumer C. *Organic Synthesis Based on Name and Unnamed Reactions*, Pergamon, Oxford, 1994, p. 36

[477] Itoh N. and Sugasawa S. *Tetrahedron* (1957) **1**, 45–8.

[478] Kanaoka Y., Sato E., Yonemitsu O. and Ban Y. *Tetrahedron Lett.* (1964) (35), 2419–22.

[478a] Bailey D.M., DeGrazia C.G., Lape H.E. Frering R., Fort R. and Skulan T. *J. Med. Chem.* (1973) **16**, 151–6.

[479] Pictet A. and Gams A. *Ber.* (1909) **42**, 2943–62.

[480] Pictet A. and Gams A. *Ber.* (1910) **43**, 2384–403.

[481] Herz W. and Tsai L. *J. Am. Chem. Soc.* (1955) **77**, 3529.

[482] Datta N.L., Wodia M.S. and Bindra A.P. *J. Indian Chem. Soc.* (1971) **48**, 873.

[483] Vatsuro K.V. and Mischenko G.L. *The Name Reactions in Organic Chemistry*, Khimia, Moscow, 1976, pp. 59–60.

[484] Fodor G., Gal J. and Phillips B.A. *Angew. Chem. Internat. Edit.* (1972) **11**, (10), 919–20.

[485] Ramesh D. and Srinivasan M. *Synth. Commun.* (1986) **16** (12), 1523–7.

[486] Koesch C.F. and Lindquist R.M. *J. Org. Chem.* (1956) **21**, 657–9.

[487] Schindlbauer H. *Monatsh.* (1968) **99**, 1799.

[488] Pizey S.S. *Synthetic Reagents. Vol. 1*, Wiley, New York, 1974, p. 74.

[489] Yassin F.A., Ahmed G.A. and Mostafa H.Y. *J. Indian Chem. Soc.* (1996) **73** (11), 620–2.

[490] Cook J.W. *J. Chem. Soc.* (1932) 1472–84.

[490a] Sandin R.B. and Fieser L.F. *J. Am. Chem. Soc.* (1940) **60**, 3098–102.

[491] Bergmann E. and Szmuszkovicz J. *Bull. Soc. Chim. France* (1953) **20**, 566–8.

[492] Middleton W.J. *J. Org. Chem.* (1965) **30** (5), 1402–7.

[493] Parkash R., Meena M. and Singh S. *Bull. Chem. Soc. Japan* (1991) **64**, 1443–4.

[494] Gramstead T. and Hazeldine R.N. *J. Chem. Soc.* (1957) 2640–5.

[495] Sokol'ski G.A., Ragulin L.I., Ovsyannikov G.P., Knunyants I.L., *Izv. Acad. Sci. USSR Chem. Sci.* (1971) 1270–5.

[496] Knunyants I.L. and Fokin A.V. *Dokl. Akad. Nauk USSR* (1957) **112**, 67–9.

[497] Knunyants I.L. and Fokin A.V. *Izv. Akad. Nauk SSSR, Ser. Khim.* (1957) 1439–51.

[498] Field L. and McFarland J.W. *Org. Synth. Collect. Vol. 4*, Wiley, New York, 1963, pp. 940–2

[499] Stang P.J. and Deuber T.E. *Org. Synth.* (1974) **54**, 79–84.

[500] Kazmierczak F. and Helquist P. *J. Org. Chem.* (1989) **54** (16), 3988–92.

[501] Pat. 07,238,053. Japan (1994).

[502] Kim J.-K., Kwon P.-S., Kwon T.-W., Chung S.-K. and Lee J.-W. *Synth. Commun.* (1996) **26** (3), 535–42.

[503] Sekera A. and Rumpf M.P. *Compt. Rend.* (1965) **260**, 2252–4.

[504] Eaton P.E. Carlson G.R. and Lee J.T. *J. Org. Chem.* (1973) **38** (23), 4071–3.

[505] Ueda M. *Org. Chem. Jap.* (1990) **48**, 144–56.

[506] Eaton P.E. and Mueller R.H. *J. Am. Chem. Soc.* (1972) **94** (3), 1014–6.

[507] Ueda M., Sato M. and Mochizuki A. *Makromolec.* (1985) **18**, 2723–6.

[508] Ueda M., Yokote S. and Sato M. *Polymer J.* (1986) **18** (2), 117–22.

[509] Ueda M., Sugita H. and Sato M. *J. Polymer Sci.* (1986) **24**, 1019–26.

[510] Stott P.E., Bradshaw J.S., Parish W.W. and Cooper J.W. *J. Org. Chem.* (1980) **45**, 4716–20.

[510a] Parish W.W., Stott P.E. and McCausland C.W. *J. Org. Chem.* (1978) **43** (24), 4577–81.

[511] Ueda M., Kano T., Waragai T. and Sugita H. *Makromol. Chem. Rapid Commun.* (1985) **6**, 847–50.

[512] Ueda M. and Kano T. *Makromol. Chem. Rapid Commun.* (1985) **5**, 833–6.

[513] Ueda M. and Sato M. *Makromol.* (1987) **20**, 2675–8.

[514] Ueda M. and Oda M. *Polymer J.* (1989) **21** (9), 673–9.

[515] Ueda M., Sugita H. and Waragai T. *Polymer J.* (1988) **20** (5), 433–7.

[516] Yonezawa N., Tokita Y., Hino T., Nakamura H. and Katakai R. *J. Org. Chem.* (1996) **61** (10), 3551–2.

[517] Popp F.D. and McEwen W.E. *Chem. Rev.* (1958) **58** (2), 321–401.

[518] Raphael R.A., Taylor, E.C. and Wynberg, H. Advances in Organic Chemistry, Vol 1. in *Polyphosphoric Acid as a Reagent in Organic Chemistry* Eds F. Uhlig and S.R. Snyder, Interscience, 1960, pp. 35–81.

[519] Bader A.R. and Kontowicz A.D. *J. Am. Chem. Soc.* (1953) **75**, 5416–7.

[520] Klibansky Y. and Ginsburg D. *J. Chem. Soc.* (1957) 1293–9.

[521] Brockmann H. and Maxfeldt H. *Chem. Ber.* (1956) **89**, 1397–402.

[522] Satchell D.P.N. and Satchell R.S. *Chem. Soc. Rev.* (1975) **4** (2), 231–50.

[523] Eleev A.F., Knuniants I.L., Pletniov S.I. and Sokolski G.A. Pat. 598,885. USSR (1978).

[524] Zavlin P.M. and Efremov D.A. *Zh. Obsch. Khim.* (1988) **58** (10), 2403–4.

[525] Zavlin P.M., Efremov D.A. and Essentseva N.S. *Zh. Obsch. Khim.* (1991) **61** (5), 1269–70.

[526] Perekalin V.V., Lipina E.S., Berestovitskaya V.M. and Efremov D.A. *Nitroalkenes*, Wiley, UK, 1994, pp. 26–7.

[527] Firl J. and Runge W. *Naturforsch.* (1974) **29**, 393–8.

[528] Fishbein P.L. and Moore H.W. *J. Org. Chem.* (1984) **49** (12), 2190–4.

[529] Moore H.W., Hernandez L. and Sing A. *J. Am. Chem. Soc.* (1976) **98** (12), 3728–30.

[530] Kunert D.M., Chambers R., Mercer F., Hernandez L. and Moore H.W. *Tetrahedron Lett.* (1978) 929–32.

[531] Chow K., Nguen N.V. and Moore H.W. *J. Org. Chem.* (1990) **55** (12), 3876–80.

[532] Tidwell T.T. *Ketenes*, Wiley, Chichester, 1995, pp. 196–9.

[533] Seikaly H.R. and Tidwell T.T. *Tetrahedron.* (1986) **41**, 2587–613.

[534] Huang W., Fang D., Temple K. and Tidwell T.T. *J. Am. Chem. Soc.* (1997) **119** (12), 2832–8.

[535] Ito T., Aoyama T. and Shiori T. *Tetrahedron Lett.* (1993) 6583–6.

[536] Takaoka K., Aoyama T. and Siori T. *Tetrahedron Lett.* (1996) 4973–6.

[537] Takaoka K., Aoyama T. and Siori T. *Tetrahedron Lett.* (1996) 4977–8.

[538] Matsumoto T., Takaoka K., Aoyama T. and Shiori T. *Tetrahedron.* (1997) **53**, 225–36.

[539] Bock H., Hirabayashi T. and Mohmand S. *Chem. Ber.* (1981) **114**, 2595–608.

[539b] Moloney D.W.J., Wong M.W., Flammang R. and Wentrup C. *J. Org. Chem.* (1997) **62** (13), 4240–7.

[540] Erickson J.L. and Kitchens G.C. *J. Org. Chem.* (1962) **27** (2), 460–3.

[541] Hazek R.H., Clark R.D., Elam E.U. and Nations R.G. *J. Org. Chem.* (1962) **27** (6), 3106–11.

[542] Baeyer A. *Ber.* (1885) **18**, 3454–61.

[543] Leuchs H. and Geserick A. *Ber.* (1908) **41**, 4171–86.

[544] Combes M.A. *Bull. Soc. Chim. Paris* (1894) **11**, 710–7.

[545] Bamfield P. and Gordon P.F. *Chem. Soc. Rev.* (1984) **13**, 441–87.

[546] Yurieva L.P. *Uspekhi Khim.* (1974) **43** (1), 95–133.

[547] Cho H., Mizuno A., Hamaguchi M. and Tatsuoka T. Pat. 0525,128 Japan (1993). *Chem. Abstr.* (1993) **119**, 95364.

[548] Cho H. and Matsuki S. *Heterocycles* (1996) **43** (1), 127–31.

[549] Walker G.N. *J. Am. Chem. Soc.* (1953) **75**, 3387–90.

[550] Horning E.C., Koo J. and Walker G.N. *J. Am. Chem. Soc.* (1951) **73**, 5826–8.

[551] Krespan C.G. *J. Org. Chem.* (1979) **44** (26), 4924–9.
[552] Alcazar V., Tapia I. and Moran J.R. *Tetrahedron* (1990) **46**, 1057–62.
[553] Saunders J.H. *Org. Synth. Collect. Vol. 3*, Wiley, New York, 1955, pp. 22–3.
[554] Condon F.E. and West D.L. *J. Org. Chem.* (1980) **45** (10), 2006–9.
[555] Condon F.E. and Mitchell G. *J. Org. Chem.* (1980) **45** (10), 2009–10.
[556] Weigert F.J. *J. Fluorine Chem.* (1971) **1**, 445–62.
[556a] Fieser M. *Reagents For Organic Synthesis. Vol. 10*, Wiley-Interscience, New York, 1982, pp. 319–20.
[557] Yu H. and Kayser M.M. *Synth. Commun.* (1994) **24**, 1583–8.
[557a] Dumontet J. *Hebd. Seances Acad. Sci.* (1952) **234**, 1173.
[558] Kaz'mina N.B., Knunyants I.L., Mysov E.I. and Kuz'yants G.M. *Izv. Akad. Nauk USSR Ser. Khim.* (1978) 163–71
[559] Stang P.J. and Dueber T.E. *Org. Synth.* (1974) **54**, 79–84.
[560] Bader A.R. and Hyre J.E. US Pat. 2,765,320 (1956).
[561] Thesing J. and Funk F.H. *Chem. Ber.* (1956) **89** (9), 2498–507.
[562] Feuer H. and Wyman J.E. *Chem. Ind.* (1956) 577.
[563] Anet F.A.L., Bavia P.M.G. and Dewar M.J.S. *Can. J. Chem.* (1957) **35**, 180–2.
[564] Navruzov Kh., Kuchkarov A.B. and Kurbanov F.K. *Tr. Tashk. Politekh. Inst.* (1972) **90**, 37–9. *Chem. Abstr.* (1975) **83**. 96629.
[565] Cargill R.L. and Jackson T.E. *J. Org. Chem.* (1974) **38**, 2125–9.
[566] Imamoto T., Matsumoto T., Kusumoto T. and Yokoyama M.A. *Synthesis* (1983) 6, 460–1.
[567] Stone H. and Shechter H. *Org. Synth. Collect. Vol. 4*, Wiley, New York, 1963, pp. 323–4.
[568] Stone H. and Shechter H. *Org. Synth. Collect. Vol. 4*, Wiley, New York, 1963, pp. 321–2.
[569] Stone, H. and Shechter, H. *Org. Synth. Collect. Vol. 4*, Wiley, New York, 1963, pp. 543–4.
[570a] Kanoaka Y., Kuga T. and Tanizawak K. *Chem. Pharm. Bull. Japan* (1970) **18**, 397.
[570b] Kanoaka Y., Hamada T. and Yonemitsu O. *Chem. Pharm. Bull. Japan* (1970) **18**, 587–90.
[570c] Kanaoka, Y., Sato E. Yonemitsu O. and Ban Y. *Tetrahedron Lett* (1964) **35**, 2419–22.
[570d] Bhati A. and Kale N. *Angew. Chem. Int. Ed.* (1967) **6**, 1086–7.
[570e] Badertscher D.E., Coonradt H.L. and Crowley J.J. US Pat. 2,386,769 (1945).
[571] Fujita S., Koyama K. and Imagaki Y. *Synthesis* (1982) 68.
[571a] Imamoto T., Yokoyama H. and Yokoyama M. *Tetrahedron Lett.* (1981) 1803–4.
[572] Kanaoka Y., Hamada T. and Yonemitsu O. *Chem. Pharm. Bull. Japan* (1970) **18**, 587–90.
[573] Yokoyama M., Yoshida S. and Imamoto T. *Synthesis* (1982) 591–2.
[574] Edmundson R.S. *A Specialist Periodical Report on Organophosphorus Chemistry* (1987) **8**, 134–86.
[575] Imamoto T., Yokoyama H. and Yokoyama M. *Tetrahedron Lett.* (1982) 1467–70.
[576] Masamune S., Hurama M., Mory S., Ali S.A. and Garvey D.S. *J. Am. Chem. Soc.* (1981) **103** (6), 1568–71.
[577] Masamune S., Ellingboe J.W. and Chov W. *J. Am. Chem. Soc.* (1982) **104** (20), 5526–8.
[578] Masamune S., Imperiali B. and Garvey D.S. *J. Am. Chem. Soc.* (1982) **104** (20), 5528–31.
[579] Bartlett P.A. *Tetrahedron.* (1980) **36**, 3–72.

[580] Stoyanovich F.M. and Marakatkina M.A. *Izv. Akad. Nauk USSR. Ser. Khim.* (1978) 1, 142–50.

[581] Gulick A. *Ann. New York. Acad. Sci.* (1957) **69**, 309–13.

[582] Miller S.L. and Urey H.C. *Science* (1959) **130**, 245–51.

[583] Miller S.L. and Parris M. *Nature* (1964) **204**, 1248–50.

[584] Yamagata Y., Matsukawa T., Mohri T. and Ionomata K. *Nature* (1979) **282**, 284–6.

[585] Yamagata Y., Mohri T., Yamakoshi M. and Ionomata K. *Origins of Life* (1981) **11**, 233–5.

[586] Oana S. *Bull. Volcanologique* (1963) **24**, 48–57.
Newman W.F. and Newman M.W. *Biological Mineralization*, Wiley, New York, 1973, p. 3.

[587] Robie R.A., Hemingway B.S., Fisher J.R. *U.S. Geol. Survey Bulletin* (1978) 1452, 456.

[588] Tageeva N.V. *Dokl. Akad. Nauk USSR* (1942) **34** (4–5), 117–20.

[589] Hall R.H. and Khorana H.G. *J. Am. Chem. Soc.* (1955) **77** (7), 1871–5.

[590] Schramm G., Grotsch H. and Pollmann W. *Angew. Chem.* (1962) **74** (2), 53–9.

[591] Schramm G. and Wissmann H. *Chem. Ber.* (1958) **91**, 1073–82.

[592] Onodera K., Hirano S. and Fukumi H. *Agr. Biol. Chem.* (1964) **28** (3), 173–8.

[593] Seela F. and Hoa Tran Thi Q. *Chem. Ber.* (1979) **112**, 3743–7.

[594] Grunza H. and Koransky W. *Angew. Chem.* (1959) **71** (12), 407–10.

[595] Fischer E. and Delbruck K. *Ber.* (1909) **42**, 2776.

[596] Micheel F., Bockman A. and Merckstroth E. *Makromol. Chem.* (1961) **48**, 1–15.

[597] Hirano S., Nishio T. and Ito T. *Arg. Biol. Chem.* (1975) **39** (10), 1963–7.

[598] Micheel F., Bockman A. and Merckstroth E. *Makromol. Chem.* (1962) **51**, 97–101.

[599] Onodera K., Hirano S. and Kashimura N. *J. Am. Chem. Soc.* (1965) **87**, 4651–2.

[600] Hirano S., Kashimura N., Kosaka N. and Onodera K. *Polymer* (1972) **13**, 190–7.

[601] Hirano S. *Agr. Biol. Chem.* (1973) **37**, 187–9.

[602] Husemann E. and Muller G.J. *Makromol. Chem.* (1961) **49**, 238–43.

[603] Husemann E. and Muller G.J. *Makromol. Chem.* (1966) **91**, 212–30.

[604] Mizuno T. *Nippon Nogeikagaku Kaishi* (1967) **41**, 189–95.

[605] Onodera K., Hirano S. and Kashimura N. *Carbohydr. Res.* (1968) **6**, 276–85.

[606] Brimacombe J.S., Bryan J.G.H., Husain A., Stacey M. and Tolley M.S. *Carbohydr. Res.* (1967) **3**, 318–24.

[607] Baker B.R. and Buss D.H. *J. Org. Chem.* (1965) **30**, 2304–8.

[608] Baker B.R. and Buss D.H. *J. Org. Chem.* (1965) **30**, 2308–11.

[609] Beynon P.J., Collins P.M., Doganges P.T. and Overend W.G. *J. Chem. Soc. (C)* (1966) 1131–6.

[610] Terner G.M. *J. Am. Chem. Soc.* (1961) **83** (1), 159–68.

[611] Peterson E.A. and Sober H.A. *J. Am. Chem. Soc.* (1954) **76** (1), 169–75.

[612] Gunsalus I.C., Umbreit W.W., Bellamy W.D. and Foust C.E. *J Biol. Chem.* (1945) **161**, 743.

[613] Heyl D., Luz D., Harris S.A. and Folkers K. *J. Am. Chem. Soc.* (1951) **73**, 3436–7.

[614] Viscontini M., Ebnother C. and Karrer P. *Helv. Chim. Acta.* (1951) **34**, 2199.

[615] Viscontini M., Ebnover C. and Karrer P. *Helv. Chim. Acta.* (1951) **34**, 1834–42.

[616] Viscontini M. and Karrer P. *Helv. Chim. Acta.* (1952) **35**, 1924–5.

[617] Baddiley J. and Mathias A.P. *J. Chem. Soc.* (1952) 2583–91.

[618] Peterson E.A. and Sober H.A. *J. Am. Chem. Soc.* (1954) **76** (1), 169–75.

[619] Schorre G. US Pat. 3,124,587 (1964).

[620] DeClercq P.J. *Chem. Rev.* (1997) **97** (6), 1755–92.

[621] Alcazar V., Tapia I. and Moran J.R. *Tetrahedron* (1990) **46** (3), 1057–62.

[622] Schul W. and Paust J. Pat. 3,906,633. Ger. (1990).

[623] Kraszewski A. and Stawinski J. *Tetrahedron Lett.* (1980) 2935–6.

[624] Knight D.W. and Pattenden G. *J. Chem. Soc. Chem. Commun.* (1976) 660–1.

[625] Fieser M. and Fieser L.F. *Reagents for Organic Synthesis. Vol. 7*, Wiley-Interscience, New York, 1979, p. 291.

[626] Ferrel R., Olcott H.S. and Fraenkel-Conrat H. *J. Am. Chem. Soc.* (1948) **70**, 2101–7.

[627] Snyder H.R. and Werber F.X. *J. Am. Chem. Soc.* (1950) **72**, 2962–5.

[628] Galat A. *J. Am. Chem. Soc.* (1951) **73**, 3654–6.

[628a] Khorana H.G. *The Enzymes. Vol. 5*, Academic Press, New York, 1961, p. 79.

[629] Bredereck H. and Berger E. *Ber.* (1940) **73**, 269–74.

[629a] Goldberg A.A. *Chem. Prod. Chem. News* (1954) **17**, 371–4.

[630] Zervas L. *J. Am. Chem. Soc.* (1955) **77**, 5354–5.

[631] Montgomery H.A.C. and Turnbull J.H. *J. Chem. Soc.* (1958) 1963–8.

[631a] Markowska A., Olejnik J., Mlotkowska B. and Sobanska M. *Phosphorus Sulphur* (1981) **10** (2), 143–6.
Markowska A., Olejnik J. and Michalski J. *Bull. Acad. Pol. Sci. Chem.* (1979) **27** (2), 115–9. Mazkowska A. and Olejnik J. Pat. 91, 923 Pol. (1997).

[632] Michelson A.M. and Todd A.R. *J. Chem. Soc.* (1949) 2476–7.

[633] Zavlin P.M. and Sokolovski M.A. *Zh. Obsch. Khim.* (1960) **30**, 3562–5.

[634] Zavlin P.M., Abuladze G.V., Efremov D.A., Chuchulashvili N.A., Gotsiridze I.A. and Govorkov A.L. Pat. 1,768,607 USSR (1992).

[635] Birch A.J. and Subba Rao G.S.R. *Tetrahedron Lett.* (1967) 29, 2763–5.

[636] Aaronson A.M. "Phosphorus Flame Retardants for a Changing World" in *Phosphorus Chemistry. Development in American Science.* ed. E.N. Walsh, E.J. Griffith, R.W. Parry and L.D. Quin. ACS Symposium Series, 486, ACS, Washington, DC, 1992, pp. 218–28.

[637] Cunnion J. *Specialty Chemicals* (1997) **17** (7), 266–7.

[638] Purdela D. and Vilceanu R. *Chemistry of Organophosphorus Compounds.* Khimia, Moscow, 1972, 22, 477–84.

[639] Hayashi K. *Makromol. Chem.* (1978) **179**, 1753–63.

[640] Hall R.E. and Munter C.J. US Pat. 2,426,394 (1947).

[641] Pincus A.G. US Pat. 2,434,673 (1948).

[642] King C.S. US Pat. 2,395,126 (1946).

[643] Hatch G.B. Pat. 449,983 Canada (1948).

[644] Tayler E.A. US Pat. 2,165,948 (1939).

[645] Vogelsang P.G., Jr. US Pat. 3,462,365 (1969).

[646] Shearon W.H., Jr. and Hoiberg A.J. *Ind. Eng. Chem.* (1953) **45** (10), 2122–32.

[647] Hoiberg A. J. U.S. Pat. 2,450,756 (1948).

[648] Toy A.D.F. *J. Am. Chem. Soc.* (1948) **7** (1), 186–8.

[649] Taber D.F., Amedio J.C. and Jung K.-Y. *J. Org. Chem.* (1987) **52** (25), 5621–2.

[650] Palomo C. and Odriozola B. *J. Chem. Soc. Chem. Commun.* (1994) 1505–7.

[651] Ai M. *Appl. Catal.* (1986) **28** (1–2), 223–33.

[652] Yui T., Ogawa H and Kitahare M., Nippon Kagaku Kaishi 481,489 (1975) *Chem. Abstr.* (1975) **83**, 59321.

[653] Friedrichsen W. and Goehre O. Ger. Patent 2,159,442 (1973) *Chem. Abstr.* (1973) **79**, 79459.

[654] McCoubrey J.C. "Elemental Phosphorus, Industrial Phosphoric acid, and their Derivatives" in *Industrial Inorganic Chemistry*, 1995, pp. 373–403.

[655] Kapishev T.A. and Menlibaev A. *Izv. Akad. Nauk Russia. Ser. Khim.* (1992) 5, 6–10.

[656] Stinson I.M., Striplin M.M. and Brown N.A. *J. Agric. Food Chem.* (1956) **4**, 248–54.

[657] Such J.E. *Lower and Higher Phosphorus Oxides. Mellor's Comprehensive Treatise on Inorganic and Theoretical Chemistry. Vol. 8. Suppl. III*, Longmans, London, 1971.

[658] Andress K.R., Fischer K. and Gehring W. *Z. Anorg. Allg. Chem.* (1949) **260**, 331–6.

[659] Born I. *Angew. Chem.* (1955) **67**, 409.

[660] Mooney R.W. and Alia M.A. *Chem. Rev.* (1961) **61**, 433–62.

[661] Van Wazer J.R. and Callis C.F. *Chem. Rev.* (1958) **58**, 1011–46.

[662] Rhône-Poulenc Technical Bulletin, 1997.

[663] Schmidlin J. and Massini P. *Chem. Ber.* (1910) **43**, 1162–71.

[664] Toennies G. *J. Am. Chem. Soc.* (1937) **59**, 555–7.

[665] Thomas P.J. and Pews R.G. US Pat. 5,120,884.

[666] Schenk P.W. and Riehag H. *Z. Anorg. Chem.* (1937) **233**, 403–10.

[667] Schenk P.W. and Vietzke H. *Z. Anorg Chem.* (1963) **326**, 152–69.

[668] Mellor J.W. *Comprehensive Treatise on Inorganic and Theoretical Chemistry, Vol. 8. Supplement III*, Wiley-Interscience, 1971, Ch. 10, pp. 404–5.

[669] Hayek E., Aignesberger A. and Englebrecht A. *Monatsh* (1955) **86**, 735–40.

[670] Tomomatsu H. U.S. Pat. 3,879,529 (1975).

[671] Lange W. *Chem. Ber.* (1929) **62**, 786–92.

[672] Hoeflake J.M.A. *Rec Trav. Chim.* (1929) **48**, 973–8.

[673] Tarbutton G., Egan E.P. and Frary S.G. *J. Am. Chem. Soc.* (1941) **63**, 1782–9.

[674] Stoddart E.M. *J. Chem. Soc.* (1945) 448–51.

[675] Stoddart E.M. *J. Chem. Soc.* (1938) 1459–61.

[676] Seel F., Schmutzler R. and Wesem K. *Angew. Chem.* (1959) **71**, 340.

[677] Partington J. R. *Textbook of Inorganic Chemistry*, Macmillan, 1961, p. 580.

[678] Osadchaya L.I., Mamaev S.A. and Serenko T.A. Pat. 1,678,760 USSR.

[679] Jansen M., Begemann B. and Geb J. *Z. Anorg. Allg. Chem.* (1992) **610**, 139–40.

[680] Buck A.C., Bartleson J.D. and Lankelma H.P. *J. Am. Chem. Soc.* (1948) **70**, 744–6.

[681] Forthmann R. and Schneider A. *Naturwiss.* (1965) **52**, 390–1.

[682] Clever H.L., Westrum E.F. and Cordes A.W. *J. Phys. Chem.* (1965) **69**, 1214–9.

[683] Brasseur P. *Bull. Soc. Chim.* (1945) **12**, 412–30.

[684] Brasseur P. *Bull. Soc. Chim.* (1946) **13**, 261–5.

[685] Shevchenko N.P., Ulanova N.M. and Serazetdinov D.Z. *Izv. Akad. Nauk Resp. Kaz. Ser. Khim.* (1991) 3–6. *Chem. Abstr.* (1993) **119**, 19301.

[686] Shevchenko, N.P., Ulanova, N.M., Serazetdinov D.Z. and Ugryumova K.E. *Izv. Akad. Nauk Resp. Kaz. Ser. Khim.* (1992) 3–9. *Chem. Abstr.* (1992) **117**, 51813.

[687] Shevchenko N.P., Ulanova N.M. and Serazetdinov D.Z. *Izv. Akad. Nauk Resp. Kaz. Ser. Khim.* (1991) 3–6. *Chem. Abstr.* (1993) **119**, 19301.

[688] Pfrengle O. Pat. 1,034,964 Ger (1958).

[689] Riou A., Cudennec Y. and Gerault Y. *Acta Crystallogr. Sect. Cryst Struct. Commun.* (1987) **43**, 194–7.

[690] Watanabe A. Jpn. Kokai Tokkyo Koho JP 02,302,321. *Chem. Abstr.* (1991) **115**, 52864.

[691] Sazhenkov A. Yu., Masloboev V.A. and Lokshin E.P. *Zh. Neorg. Khim.* (1993) **38**, 940–2.

[692] Nagorni P.G. and Kapshuk A.A. *Zh. Neorg. Khim.* (1993) **38**, 11–13.

[693] Li K.H., Shih P.F. and Chen T.M. *Inorg. Chem.* (1993) **32**, 4372–7.

[694] Lu C. and Lii K.H. *J. Solid State Chem.* (1991) **92**, 362–9.

[695] Hwu S.J. and Willis E.D. *J. Solid State Chem.* (1991) **93**, 69–76.

[696] Lii K.H. and Tsai H.J. *J. Solid State Chem.* (1991) **90**, 291–5.

[697] Wang S. and Hwu S.J. *J. Solid State Chem.* (1991), **90**, 34–41.

[698] Hanshalter R.C. *et al. J. Solid State Chem.* (1990), **89**, 215–9.

[699] Watanabe A. and Takenochi S. Wignacourt J.P., Conflant P., Drache M. and Boivin T.C. *J. Solid State Chem.* (1993) **107**, 93.

[700] Watanabe A. and Takenochi S. Jpn. Kokai Kokkyo Koho JP 06,128,087. *Chem. Abstr.* (1994) **121**, 218325.

[701] Kryukova A.I., Masterova L. Yu and Skiba O.V. *Radiokhimika* (1990) **32**, 75–9.

[702] Juszczyk K. and Podsiadlo V. *Pr. Nauk Akad. Ekon. im. Oskara Langego Wroclawia* (1990) **526**, 231–2. *Chem. Abstr.* (1991) **115**, 104600.

[703] Belghiti A.E., Boukhari A. and Holt E.M. *J. Alloys Compds.* (1992) **188**, 128–32.

[704] Forsyth J.B., Wilkinson C., Paster S. and Wanklyn B.M. *J. Phys. Condens. Matter.* (1989) **1**, 169–78.

[705] Belghitti A.E., Boukhari A. and Holy E.M. *Acta Ctrystallogr. Crst. Struct. Commun.* (1994) **50**, 482–4.

[706] Takita Y., Yamahita H and Mritaka K. *Chem. Lett.* (1993) 335–8.

[707] Handizi A., Belghitti A.E., Boukhari A., Holt E.M., Aride J., Belaiche M. and Drillon M. *Eur. J. Solid Stat Inorg. Chem.* (1994) **31**, 123–36.

[708] Bitterwolf T.E. and Dai X. *J. Orgonamet. Chem.* (1992) **440**, 103–112.

[709] Ansorge A., Brauer D.J., Buerger H., Krumm B. and Pawelke G. *Organomet. Chem.* (1993) **446**, 25–36.

[710] Herberich G.E. and Barlage W. *J. Organomet. Chem.* (1987) **331**, 63–72.

[711] Herberich G. E. and Barlage W. *Organometallics* (1987) **6**, 1924–30.

[712] Vishnyakova T.P., Golubeva I.A. and Paushkin Ya.M. *Polym. Sci.* (1965) **7**, 787–92.

[713] Vishnyakova T.P., Golubeva I.A. and Paushkin Ya.M. *Polym. Sci.* (1966) **8**, 196–201.

[714] Graham P.J., Lindsey R.V., Parshall G.W., Peterson M.L. and Whitman G.M. *J. Am. Chem. Soc.* (1957) **79**, 3416–20.

[715] Lindsay J.K. and Hauser C.R. *J. Org. Chem.* (1957) **22**, 355–8.

[716] Mueller K., Heyns A.M., Range K-J. and Zabel M.Z. *Naturforsch. B. Chem Sci.* (1992) **47**, 238–46.

[717] Fischer D., Hoppe R., Schaefer W. and Knight K.S.Z. *Anorg. Allg. Chem.* (1993) **619**, 1419–25.

[718] Kuzina A.F., Oblova A.A. and Spitsyn V.I. *Zh. Neorg. Khim.* (1972) **17**, 2630–2.

[719] Blechschmitt K. *et al.* Ger. Offen. 2,514,989 (1976).

[720] Saito S. U.S. Pat. 3,859,326 (1974).

[721] Eden J. S. U.S. Pat. 3,708,434; 3,737,394 (1973).

[722] Ono I. Ger. Offen. 2,202,733 (1973).

[723] Durif A. *Crystal Chemistry of Condensed Phosphates*, Plenum Press, New York, 1995.

[724] Friedrichsen W. and Goehre O. Pat. 2,159,442 Ger. (1973) *Chem. Abstr.* (1973) **79**, 79459.

[725] Abe Y., Hosono H., Kamae T. and Kawashima K. *Phosphorus Sulfur* (1990) **49–50** (Part 2, 1–4), 113–6.

[726] Brosheer J.C., Tennessee Valley Authority Chem. Eng. Report No. 6. Wilson Dam, Alabama. 1953.

[727] Ger. Pat. DE 4,240,191. *Chem. Abstr.* (1994) **121**, 41295.

[728] Bourguiba N.F. and Dogguy L.S. *Mater. Res. Bull.* (1994) **29**, 427–36.

[729] Tomlinson W.F. US Pat. 2,870,091 (1959).

[730] King C.S. US Pat. 2,370,472 (1946).

[731] Mendenhall E. E. US Pat. 2,510,510 (1950).

[732] Pincus A.G. US Pat. 2,434,674 (1948).

[733] Pincus A.G. US Pat. 2,319,260 (1943).

[734] Englund L.H. US Pat. 2,375,638 (1945).

[735] Andress H.J. and Asnjian H. US Pat. appl. 907241. *Chem. Abstr.* (1992) **117**, 11331.

[736] Khisameev G.G., Sirotkin O.S. and Saitullin R.S. *Izv. Akad. Nauk USSR Neorg. Mater.* (1990) **26**, 2198–201.

[737] Fukumura C., Inine K., Iwata M. and Navita N. Eur Patt. Appl. EP 494,778. *Chem. Abstr.* (1973) **118**, 4183.

[738] Staffel T., Becker W. and Neumann H., Ger. Offen. DE 4,133,811. *Chem. Abstr.* (1993) **118**, 237073.

[739] Shearon W.H., Jr. and Hoiberg A.J. *Ind. Eng. Chem.* (1953) **45** (10), 2122–32.

[740] Hoiberg A.J. US Pat. 2,450,756 (1948).

[741] Nettles J.E. *Handbook of Chemical Specialties. Textile Fiber Processing, Preparation and Bleaching*, Wiley-Interscience, New York, 1983, pp. 183–92.

[742] Mayhew R.L. and Krupin F. *Soap Chem. Specialities* (1962) **38** (4), 55–58, 93, 95; (5), 80, 81, 167, 169.

[743] US Pat. 2,027,785.

[744] Chadwick D.H. and Watt R.S. "Manufacture of Phosphate Esters and Organic Phosphorous Compounds" in *Phosphorus and its Compounds*, ed. J.R. Van Waser, Interscience Publishers, London, 1961, pp. 1221–79.

[745] US Pat. 2,137,792.

[746] Gantz G.M. *Am. Dyestuff Reptr.* (1968) **57** (35) 886–99.

[747] Bergman C.A. and Robinson J. *J. Am. Dyestuff Reptr.* (1962) **51**, 114.

[748] Linder K. *Tenside-Textilhilfsmittel-Waschrohstoffe. Vol. 1*, Wissenschaftliche Verlagsgesellschaft, Stuttgart, 1964.

[749] Riethmayer S.A. *Sefen, Ole, Fette, Wachse* (1968) **94**, 906–13.

[750] *The Handbook of Industrial Surfactants*, compiled Ash M. and Ash I, Gower, Aldershot, 1993.

[751] Melnik A.P., Gavrilov G.V., Polkovnichenko I.T., Chaplanov P.E. and Legeza V.M. Pat. 1,310,400 USSR (1986).

[752] Gruen A. and Kade F. Pat. 240,075 Ger. (1910)

[753] Nusslein J. and Schuette H. US Pat. 1,914,331 (1933).

[754] Harris B.R. US Pat. 2,052,029 (1933).

[755] Harris B.R. US Pat. 2,177,650 (1939).

[756] Harris B.R. US Pat. 2,177,983 (1939).

[757] Burnette L.W. "Nonionics as Ionic Surfactant Intermediates" in *Nonionic Surfactants*, ed. M.J. Schick, Surfactant Science Series, Vol. 1, Dekker, New York, 1966, pp. 385–94.

[758] Burnette L.W. "Miscellaneous Nonionic Surfactants" in *Nonionic Surfactants*, ed. M.J. Schick, Surfactant Science Series. Vol. 1, pp. 417–420.

[759] Cooper R.S. *J. Am. Oil Chemist's Soc.* (1963) **40**, 642–9.

[760] Cooper R.S. and Urfer A.D. *J. Am. Oil Chemist's Soc.* (1964) **41**, 337–42.

[761] Meyhew R.L. and Krupin F. *Soap Chem. Specialities* (1962) **38** (4), 55–8.

[762] Meyhew R.L. and Krupin F. *Soap Chem. Specialities.* (1962) **38** (5), 80–6.

[763] Eisman Jr. F.S. and Schenk L.M. US Pat. 3,331,896 (1967).

[764] Nehmsmann L.J. and Schenck L.M. US Pat. 3,462,520 (1969).

[765] Stayner R.D. US Pat. 2,693,482 (1954).

[766] Kaneko T.M. US Pat. 3,741,912 (1973).

[767] Bertsch H. US Pat. 1,900,973 (1933).

[767a] Chatha M.S. *Ind. Eng. Chem. Res.* (1987) **26** (12), 2495–7.

[767b] Cooper R.S. and Urfer A.D. US Pat. 3,329,615 (1967).
[768] Ehara M. and Fujii T. Pat. 7,451,400 Japan (1974).
[769] Ionescu M., Antohe N., Murariu G., Stejaru C., Cazacu G., Suprateanu G. and Gavriliu N. Pat. 101,478 Rom. (1992).
[770] Barbera M.A. US Pat. 3,609,075 (1971).
[771] US Pat. 3,004,057 (1963).
[772] Caldwell J.R. US Pat. 3,153,080 (1964).
[773] Sonntag N.O.V. and Rini S.J. "Tests and Testing Methods" in *Fatty Acids and Their Industrial Applications*, ed. E.S. Pattison, Marcel Dekker, New York, 1968, pp. 330–1.
[774] Hochwalt C.A., Lum J.H., Malowan J.E. and Dyer C.P. *Ind. Eng. Chem.* (1942) **34**, 20.
[775] Gutmann W. and Linke W. Pat. 1,091,278 (1960).
[776] Pat. 851,842 France (1940).
[777] Pat. 310,907 Austr. (1973).
[778] Hathaway H.D. US Pat. 3,399,144 (1968).
[779] Eiseman F.S. and Schenk L.M. US Pat. 3,331,896 (1967).
[780] Mansfield R.C. and Hill C. US Pat. 3,235,627 (1966).
[781] Tanaka S. and Okayasu H. Pat. 08 59,887 Japan (1996).
[782] Nunn Jr. L.G. US Pat. 3,004,056 (1961).
[783] Ulrich H. and Sauerwein K. Pat. 696,317 Ger. (1940).
[784] Grifo R.A., Mayhew R.L., Stefcik A. and Woodward F.E. US Pat. 3,122,508 (1964).
[785] Jayne D.W., Jr. US Pat. 2,545,357 (1951).
[786] Stefcik A. and Woodward F.E. US Pat. 3,122,508 (1964).
[787] Stuart D.C. and Crandall H.W. *J. Am. Chem. Soc.* (1951) **73** (3), 1377–8.
[788] Crandall H.W. and Stewart D.C. US Pat. 2,658,909 (1953).
[789] Hochwalt C.A., Lum J.H., Malowan J.E. and Dyer C.P. *J. Ind. Chem.* (1942) **34** (1), 20–5.
[790] Maurer E.W., Stirton A.J., Ault W.C. and Weil J.K. *J. Am. Oil Chemist's Soc.* (1964) **41**, 205–11.
[791] Vogeli R.C. *Soap, Parfumery, Cosmetics* (1968) **41**, 503.
[792] Nusslein *J. Am. Perfumer Aromat.* (1960) **75** (7), 43.
[793] Kurosaki T., Furugaki H., Matsunaga A., Yuzawa M. and Mauba A. *Yukagaku.* (1990) **39** (4), 250–8.
[794] Takahashi M., Ikemori S., Fukuda K., Kamata Y., Takada M. and Miyakoshi S. Pat. 6,233,190 Japan (1987).
[795] Smith H.G., Hill M.L. and Cantrell T.L. US Pat. 2,442,024 (1948).
[796] Wasserman K.J. US Pat. 3,453,254 (1969).
[797] Clarke J.V., Cranford N.J. and Smith Jr. J.O. US Pat. 3,010,903 (1961).
[797a] Weil E.D. "Phosphorus-Based Flame Retardants" in *Handbook of Organophosphorus Chemistry.* Ed. R. Engel. Marcel Dekker. New York, 1992. pp. 683–9.
[798] Waly A., El-Aref A.T., Abdel-Mohdy F.A., Zamzam N.E. and Hebeish A. *Polymers & Polymer Composites* (1994) **2** (1), 27–34.
[799] Hargis S.R., Jr. US Pat. 3,574,794 (1971).
[800] Britton E.C., Marshall H.B. and LeFevre W.J. US Pat. 2,186,360 (1940).
[801] Weil E.D. US Pat. 3,666,712 (1972).
[802] Weil E.D. US Pat. 3,821,336 (1974).
[803] Pat. 6516345 Neth. (1966).
[803a] Aaronson A.M. and Bright D.A. *Phosphorus Sulfur* (1996) **109–10**, 83–6.
[804] Otto F.P. and Finn E.W. US Pat. 2,790,765 (1957).

[805] Horodysky A.G. US Pat. 4,681,692 (1987).

[806] Hutchinson J.F. and Schilling A.A. US Pat. 2,244,705 (1941).

[807] Caprio A.F. US Pat. 2,237,336 (1941).

[808] Prutton C.F. US Pat. 2,224,695 (1940). Schilling A.A. US Pat. 2,271,044 (1942).
Swezey F.H., Bishop M.E. and Hoffman R.M. US Pat. 2,278,747 (1942).
Benning A.F. US Pat. 2,167,867 (1939).
Downing F.B., Benning A.F. and Johnson F.W. US Pat. 2,285,853 (1942).
Downing F.B., Benning A.F. and Johnson F.W. US Pat. 2,285,854 (1942).
Mulit L.H. US Pat. 2.274.302 (1942).

[809] Smith H.B. US Pat. 2,021,901 (1935).
Smith H.B. US Pat. 2,021,902 (1935).
Sonneveld A. US Pat. 2,255,828 (1941).
Atkins S.A. US Pat. 2,031,629 (1936)

[810] Woodruff J.C. US Pat. 2,010,123 (1935)

[811] Grindahl G.A., Greenwald J.R., Troporcer L.H., Pierce O.R. and Kim Y.K. *J. Polym. Sci., Polym. Chem. Ed.* (1974) **12** (7), 1559–64.

[812] Fritz C.G. and Warnell J.L. US Pat. 3,347,901 (1967).

[813] Strauss W. and Heyden R. US Pat. 3,084,111 (1963).

[814] Campbell R.D., Langer H.G and Martin P.H. US Pat. 4,397,970 (1983).

[815] Paulovic M., Sutor L., Selo T., Spacir J., Novak L., Kunovsky J, Bartonicek R. and Kulhavy T. Pat. 228,816 CS (1986).

[816] Kobayashi T., Saito K. and Suzuki Y. Pat. 0687,955 Japan (1994).

[817] Pat. 07,324,129 Japan (1995).

[818] Ranny M., Sedlacek J., Hrdina P., Kondelik P. and Stribrnsky V. Pat. 256,691 CS (1988).

[819] Mc Gregor R.R. and Warrick E.L. US Pat. 2,459,387 (1949).

[820] Hyde J.F. US Pat. 2,571,039 (1951).

[821] Pat. 1,036,694 France (1953).

[822] Pat. 687,759. Brit. (1953).

[822a] Mueller F.W.H. US Pat. 2,420,610 (1947).

[823] Zavlin P.M., Dyakonov A.N., Kuznetsov L.L. and Efremov D.A. *Proceedings of IS & T's*, Rochester, 1994, pp. 257–60.

[824] Zavlin P.M., Dyakonov A.N., Kuznetsov L.L. and Efremov D.A. *Zh. Nauchn. Prikl. Photogr.* (1995) **40** (1), 51–6.

[825] Zavlin P.M., Diakonov A.N., Kuznetsov L.L. and Efremov D.A. *Proceedings of IS&T's 48th Annual Conf.*, Washington, 1995, pp. 195–6.

[826] Efremov D.A., Gugkaev A.V. and Zavlin P.M. *Proc. Intern. Congr. Photogr. Sci. (ICPS' 90)*, Beijing, 1990, pp. 42–4.

[827] Zavlin P.M., Breslav Yu.A., Efremov D.A., Gugkaev A.V., Sechkarev B.A., Spirina Yu.R. and Terentev E.G. Pat. 1,616,101 USSR (1990).

[828] Claes F.H. and Leber F. *J. Phot. Sci.* (1973) **21**, 39–50.

[829] Claes F.H. and Leber F. *J. Phot. Sci.* (1976) **24** (2), 51–6.

[830] Maskasky J.E. *J. Imag. Sci.* (1986) **30** (6), 247–54.

[831] Maskasky J.E. Europ. Pat. 213,963 (1987).

[832] Maskasky J.E. Europ. Pat. 215,612 (1987).

[833] Maskasky J.E. Europ. Pat. 213,964 (1987).

[834] US Pat. 4,724,200 (1988).

[835] US Pat. 4,680,255 (1987).

[836] Zavlin P.M., Efremov D.A., Govorkov A.L. and Essentseva N.S. *Photochim. Photophys. Proc. Silver Halides. Vsesouzn. Symp.*, (USSR) Chernogolovka, 1991, Abstracts, p. 6.

[837] Hagemeyer Jr. H.J. US Pat. 2,596,679 (1952).
[838] Hassner A. and Stumer C. *Organic Synthesis Based on Name and Unnamed Reactions*, Pergamon, Oxford, 1994, p. 252.
[839] Noller C.R. and Dutton G.R. *J. Am. Chem. Soc.* (1933) **55**, 424–5.
[840] Anonymous. *Chem. Eng. News.* (1956) **34**, 2590–2.
[841] Anonymous. *Chem. Eng.* (1956) **63** (5), 130, 132, 134.
[842] Anonymous. *Chem. Eng.* (1957) **64** (4), 149–150.
[843] Bailey G.C. US Pat. 2,409,727 (1946).

Index